The Institute of Biology's
Studies in Biology no. 9

The Electron Microscope in Biology

B 94

by A. V. Grimstone M.A., Ph.D.

*Assistant Director of Research,
Department of Zoology, University of Cambridge*

Edward Arnold (Publishers) Ltd

First published 1968
Reprinted 1970

Boards Edition SBN: 7131 2185 8

Paper Edition   SBN: 7131 2186 6

Printed in Great Britain by
William Clowes and Sons Ltd, London and Beccles

General Preface to the Series

It is no longer possible for one textbook to cover the whole field of Biology and to remain sufficiently up to date. At the same time students at school, and indeed those in their first year at universities, must be contemporary in their biological outlook and know where the most important developments are taking place.

The Biological Education Committee, set up jointly by the Royal Society and the Institute of Biology, is sponsoring, therefore, the production of a series of booklets dealing with limited biological topics in which recent progress has been most rapid and important.

A feature of the series is that the booklets indicate as clearly as possible the methods that have been employed in elucidating the problems with which they deal. There are suggestions for practical work for the student which should form a sound scientific basis for his understanding.

1968

INSTITUTE OF BIOLOGY
41 Queen's Gate
London, S.W.7.

Preface

The electron microscope is currently one of the most important tools of biological research. This booklet describes briefly how it works, and gives an outline of the information which it has yielded so far about the structure and function of living organisms. Electron microscopes are complicated and expensive instruments, but the basic principles involved are not difficult to grasp. I have tried to explain them in as simple a form as possible. The methods used in preparing material for examination are described in detail, since without understanding these it is difficult for the reader to interpret what he sees when he looks at an electronmicrograph. The limitations, as well as the potentialities, of the electron microscope, are emphasized.

I am grateful for permission to reproduce certain illustrations, and a full acknowledgement is included in the relevant captions.

Cambridge, 1968   A. V. G.

Contents

1 The electron microscope   1

    1.1 Introduction   1
    1.2 Resolving power   1
    1.3 Wave properties of electrons   6
    1.4 Electron guns   6
    1.5 Electron lenses   8
    1.6 Simple electron microscopes   8
    1.7 Modern electron microscopes   10
    1.8 Other types of electron microscope   13

2 Viruses and molecules   15

    2.1 Introduction   15
    2.2 Preparative techniques for viruses   16
    2.3 Viruses   20
    2.4 Molecules   24

3 Cells and tissues   26

    3.1 Techniques of preparation   26
    3.2 An outline of cell fine structure   32
    3.3 The problem of artefacts   38
    3.4 Cell structure and cell function   41
    3.5 Bacterial cells   45
    3.6 Specializations in cells and tissues   46

4 Surfaces   49

    4.1 Replicas   49
    4.2 Plant and animal surfaces   50

5 Conclusions   52

Further reading   54

The Electron Microscope 1

1.1 Introduction

We derive our information about the world, in the first instance, mainly by looking at it. Sight is our dominant sense, and our sensory world is primarily a visual one. Magnifying glasses, microscopes and telescopes provide extensions of our visual sense, enabling us to see, or see more clearly, small or distant objects. The invention and gradual perfection of these instruments, permitting the exploration of regions of the world inaccessible to our unaided eyes, is a major and well-known theme in the history of science. Much of the progress in biology in the last hundred years or so is closely bound up with improvements in the microscope. The discovery and study of bacteria and protozoa, the identification of cells as the fundamental constituents of plants and animals, the recognition of chromosomes and their role in heredity, the gradual increase in our understanding of the structure and development of tissues and organs—all these, to mention only some of the more obvious examples, were made possible only by the development of ever more perfect microscopes and the associated techniques of specimen preparation.

The electron microscope is one of the most recent stages in this development. It renders open to direct inspection a level of structure finer than any accessible before, and, just as did the light microscope, it reveals a new world, many of the features of which were previously unsuspected. As a result of the new observations which it has made possible, our understanding of the organization of plant and animal tissues has been enormously extended, and many of our ideas about the way cells are constructed and the way they function have been radically changed. The electron microscope has also added greatly to our knowledge of the structure and reproduction of viruses.

This booklet sets out to explain briefly the principles on which the electron microscope works, and to describe some of the biological discoveries which have been made with it.

1.2 Resolving power

In order to grasp the reasons which led to the development of the electron microscope it is first necessary to understand some of the limitations of the light microscope. This necessitates a rudimentary knowledge of microscope optics, and in particular of the meaning of the term *resolving power*.

The type of microscope with which the reader is most likely to be familiar is the ordinary compound microscope which uses visible light. This is referred to as an 'ordinary' microscope because there are several other kinds of microscope which also use visible light but are optically more complicated; the phase-contrast, interference and polarizing microscopes are the main examples. The ordinary light microscope—which will simply be called the light microscope from now on—consists essentially of a light source and three sets of lenses. The *condenser* focuses light on to the object, a magnified image of which is formed by the combined action of the *objective* and *eyepiece*, as shown in Fig. 1–1.

In constructing the path of light rays through an optical system, as in Fig. 1–1, use is made of what are called ray optics. That is, it is assumed that light travels in straight lines and undergoes refraction on entering a medium of different refractive index. Using ray optics it is possible, knowing the focal length of the various lenses and their mutual arrangement, to

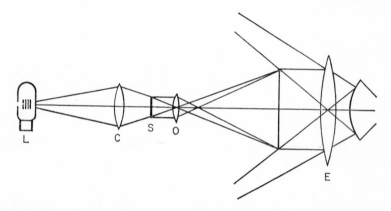

Fig. 1–1   Diagram showing in simplified form the path of rays through the light microscope. Light from the lamp (L) is focused on to the specimen (S) by the condenser (C). The image formed by the objective (O) is further magnified by the eyepiece (E).

calculate the magnification of a microscope. This can be expressed, for present purposes, as a linear magnification, which is the ratio of the length of the final image to that of the object. The magnification of light microscopes usually lies between about ×25 at the lower end of the range, to × 1,500 at the upper.

The magnification of a microscope is easy to calculate but it is not, in itself, a particularly useful piece of information. The statement that the magnification of a microscope, using a certain combination of lenses, is say, × 400, tells us by how much the apparent size of the object will have been increased, but gives no information about a much more important characteristic, namely the amount of detail we may expect to be able to see. Our aim in using a microscope is, of course, to see more detail than we can make out with our unaided eyes. Magnification is only a means to this end, and it is of no value if the image finally produced contains no more detail than we could see—or *resolve*—by eye.

The fineness of detail which can be made out with a microscope is referred to as its *resolving power*. Suppose we are examining two small objects. Provided they are well separated from each other a microscope will form distinct images of them, but if they are gradually brought closer together a stage will eventually be reached at which the two images merge and the microscope will fail to distinguish them as separate. It is in these terms that resolving power can be defined: it is the smallest separation at which we can see that two objects are present rather than one. The smaller the resolving power, the greater the amount of detail that can be made out.

The factors which limit the resolving power of a microscope cannot be understood in terms of ray optics or the kind of diagram shown in Fig. 1–1. It is necessary instead to take account of the fact that light has wave properties, and to consider the way in which the image in a microscope is formed. What happens to light when it encounters an object in the microscope? This is a rather complicated topic, the theory of which was worked out in the second half of the nineteenth century, largely by Ernst Abbe. It is unnecessary to go into this theory in any detail here, but one or two basic points must be brought out, since they are relevant to an understanding of the electron microscope.

An object becomes visible in a microscope as a result of an interaction between it and the light waves used to illuminate it. This interaction, which causes a disturbance or deviation of the waves as they pass the object, is called diffraction. Light waves which do not interact with the object will not be diffracted. According to Abbe's theory, the detail in the final image arises as a result of interference between the diffracted and undiffracted light, which in some parts of the image plane will arrive in phase with each other and tend to reinforce, while in others they will arrive out of phase and tend to cancel each other out. This results in a pattern of light and dark areas, which is the image of the object. For a detailed exposition of this theory the reader must refer to books on the theory of the microscope (see Further Reading, p. 54). For present purposes all we need do is consider in slightly more detail the question of the interaction between light waves and object. The essential point here is that light waves will be disturbed only by an object which is sufficiently large in relation to their wavelength. Very small objects will not bring about any detectable deviation in the waves, and will therefore remain invisible (or unresolved). It is not difficult to see that the smaller the wavelength of the light the smaller the object can be which will cause diffraction, and hence the better the resolving power will be. The resolving power of a microscope is, in fact, directly related to the wavelength of light used to illuminate the object.

This is an extremely important result. It means that the nature of light itself sets a limit to the amount of detail that can be resolved in a microscope. The reason why the upper limit in magnification of a microscope is set at about × 1,500 is that at that level we can comfortably see all the detail that we can hope to make out. Further magnification merely results in a larger, but no more informative image, and is analogous to examining a photograph in a newspaper with a magnifying glass in the hope of making out more detail. In the microscope it is the wavelength of light which limits the amount of detail; in the printed picture it is the number of dots to the square inch.

The explanation of image formation in terms of diffraction of light waves leads to another important result, which cannot be explained without giving a detailed account of the whole theory, and can only be stated here in general terms. It is that the resolving power of a microscope will depend not only on the wavelength of the illumination but also on the extent to which diffracted light is accepted by the objective lens. This in turn is related to the angle subtended at the front lens of the objective by the

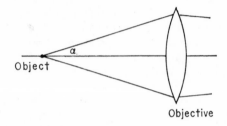

Fig. 1–2 Diagram defining the semi-angle (α) subtended by the object at the objective lens.

object (Fig. 1–2). If this angle is small (as, for example, when the objective is far away from the specimen) then only a narrow cone of light is accepted by the lens and much of the diffracted light is lost. The resolving power is consequently poor. Conversely, when the angle is large, more of the diffracted light is admitted and the resolving power is improved. It is for this reason that high-power objective lenses have to be brought up very close to the specimen, while low-power ones have a large working distance.

Taking into account the two factors, wavelength (λ) and the semi-angle α subtended by the object at the objective, the resolving power (δ) can be expressed as a first approximation by the equation:

$$\delta = \frac{\lambda}{\sin \alpha}$$

To be exact, however, two further factors need to be taken into account. Firstly, the resolving power depends not only on the angle α but also on the refractive index (n) of the medium between the specimen and the objective. In most cases the medium is air, which leaves the equation unchanged, but the highest power objectives are used with oil or water filling the space between objective and specimen, and the resolving power is then reduced by $1/n$. In the case of oil the value n is about $1\cdot4$. Secondly, for theoretical reasons which need not be entered into here, a constant, $0\cdot5$, has to be introduced into the equation, which now becomes:

$$\delta = \frac{0\cdot5\ \lambda}{n\ \sin\ \alpha}$$

The quantity $n\ \sin\ \alpha$ is termed the numerical aperture (N.A.) of the objective.

We can now insert some real values in the equation and calculate the resolving power. The unit of measurement most commonly used in light microscopy is the *micron* (μ), which is 10^{-4} cm. Visible light has a wavelength of about $0\cdot5\ \mu$, and the numerical aperture of the best lenses is about $1\cdot4$. From these two values the resolving power comes out as approximately $0\cdot2\ \mu$. This is the best resolving power that can be achieved with the ordinary microscope; the numerical aperture cannot be increased, and even using visible light of the shortest possible wavelength (about $0\cdot45\ \mu$) the resolving power is not significantly improved.

It should be clear that the only way to improve resolving power is to decrease the wavelength of the illumination. To make the effect of wavelength quite clear it may be helpful to consider a simple analogy. Suppose a blindfolded person were trying to explore a room simply by feeling around with a probe. The amount of detail he could detect would depend on the thickness of the instrument he used. With a walking stick he could make out only the grosser objects—the chairs and tables, the position of the door, and so on. To resolve finer detail—to discover whether the chairs were carved or plain, for example—it would be necessary to use a finer instrument, such as a pencil. With a needle it might be possible to make out the texture of the upholstery. The thickness of the probe, which here limits resolution, is the analogue of wavelength in the microscope.

Ultraviolet light, with a wavelength of about $0\cdot3\ \mu$, offers the possibility of an approximately twofold improvement in resolving power. Unfortunately, the practical difficulties of working with ultraviolet light are fairly considerable, since it is not significantly transmitted by glass. In consequence the lenses of the microscope, as well as the slides and coverglasses, have all to be made of quartz or fluorite, which are expensive materials and difficult to work. A special lamp is, of course, necessary, and since our eyes cannot detect ultraviolet radiation (except in so far as they are damaged by it) the image has to be recorded photographically. Under the best conditions the ultraviolet microscope has a resolving power of about $0\cdot1\ \mu$, and until the advent of the electron microscope this represented the limit of resolution. Subsequently, however, it has been entirely superseded as a high-resolution instrument. Its usefulness now lies in the fact that certain cell components (such as nucleic acids) selectively absorb ultraviolet light of characteristic wavelengths, and can thereby be identified and located in the cell. The study of the localization of different chemicals in cells is called cytochemistry, and the ultraviolet microscope is still an important tool in this kind of work.

For a really significant improvement in resolving power, however, it is necessary to turn from visible or ultraviolet light to some other form of radiation, and it is here that the development of the electron microscope begins.

1.3   Wave properties of electrons

The reader was probably introduced to electrons as particulate constituents of atoms, circling the nucleus in a series of concentric orbits. While electrons can certainly be treated as particles for some purposes, from the present point of view it is more important to know that in appropriate conditions they also display wave properties. In other words, they have some of the characteristics of other forms of wave motion (such as visible light, ultraviolet light, and X-rays) and can be treated theoretically in much the same way. In particular, electrons just like visible light, have a wavelength associated with them. The wave properties of electrons were predicted on theoretical grounds by the French physicist, de Broglie, in 1924, and were confirmed experimentally a few years later.

In the next section we shall describe how beams of electrons can be produced and what factors affect their wavelength. That discussion can be anticipated, however, by saying that the wavelength of electrons is always very much less than that of visible light, and that this opens the way to a correspondingly large improvement in resolving power.

1.4   Electron guns

The source of illumination in an electron microscope is called the electron gun. This usually consists of a small V-shaped piece of wire, the filament or cathode, together with two circular metal plates with holes drilled in their centres (Fig. 1–3). A large voltage is applied between the filament (negative) and one of the plates, the anode (positive). A current flows through the filament and heats it to incandescence, causing it to emit electrons. These are attracted towards the anode, which is oppositely charged, and some of them pass through the hole in its centre. The proportion which does so is increased by the presence of the second metal plate, which is negatively charged with respect to the filament. It is called the cathode shield. Its negative charge has the effect of concentrating the electrons emitted by the filament into a beam passing symmetrically along the axis of the gun, and hence through the hole in the middle of the anode.

The property of emitting electrons when heated is common to all metals and is called *thermionic emission*. It is used in cathode-ray tubes, in valves, and in various other electronic devices. The higher the temperature of the filament the more electrons are emitted, and in practice it is usual to make the filament of tungsten, which can be heated to over 3,000 °C without melting.

The effect of the gun is to produce a narrow beam of electrons passing at high speed in a given direction. The voltage applied between the filament and the anode and which accelerates the electrons is usually between 40,000 and 100,000 volts (40–100 kV), though a few microscopes have been built which work at higher voltages than this. The greater the voltage, the

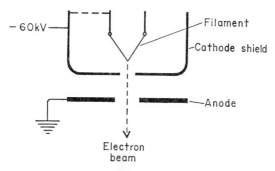

Fig. 1–3   Diagram of an electron gun.

greater the energy of the electrons. The accelerating voltage (V) also determines the wavelength of the electrons. The unit of length in which this is measured and which is commonly used in electron microscopy is not the micron (μ) but the Ångström unit (Å), which equals 10^{-8} cm ($1\ \mu = 10^4$ Å) The wavelength of electrons, in Å, is given by the approximate formula:

$$\lambda = \sqrt{\frac{150}{V}},$$

where V is in volts. Thus, using an accelerating voltage of 60 kV, the wavelength will be $\sqrt{\dfrac{150}{60,000}}$, which equals 0·05 Å. Comparison of this figure with the wavelength of visible light (0·5 μ or 5,000 Å) shows that the effective wavelength of an electron beam is about 10,000 times less than that of a light wave. Other things being equal, this would lead to a corresponding increase in resolving power. However, as we shall see, there are many factors which prevent this theoretical degree of improvement being realized in practice and the actual gain in resolving power is very much smaller. Nevertheless, although the theoretical improvement cannot be achieved, the gain in resolution made possible by the short wavelength of electrons is still very great. The actual resolving power which has been obtained will be discussed after the other components of the electron microscope have been considered.

1.5   Electron lenses

The fact than an electron beam is deflected by a magnetic field has been known since the turn of the century, and has long been made use of in the cathode-ray tube. It was not until the 1920's, however, that it was demonstrated that a radially symmetrical magnetic field, such as is formed by a coil of wire with a current passing through it, will act as a lens and can be used to focus an electron beam.

The actual path of electrons in such a lens need not be considered in detail here. It can be deduced by application of the 'left-hand' rule, more commonly applied to electric motors, and in general it is found that the path of electrons is helical. By appropriate design a lens can be produced which will form a point image of a point source of electrons. In practice

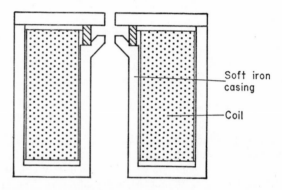

Fig. 1–4   Diagram of an electron lens in section.

an electron lens (Fig. 1–4) consists basically of a lens coil, made up of a few thousand turns of wire, with a current of about 1 amp or less flowing through it. The magnetic field produced is concentrated by a soft iron casing around the coil and, in some cases, by other, specially shaped pieces of soft iron, called polepieces, in the centre of the coil. The focal length of such a lens depends on the value of the current flowing through it, and this can usually be varied. At maximum excitation of the coil the focal length of the lenses used in an electron microscope is usually a few millimetres.

1.6   Simple electron microscopes

The electron gun and the electron lens just described together form the essential components of an electron microscope. In its general layout this is basically similar to a light microscope, except that it is inverted. A simple version is illustrated in Fig. 1–5. The electron gun takes the place of the lamp and acts as the source of illumination, and the glass lenses of the light microscope are replaced by electron lenses of similar functions. Our eyes are not, of course, sensitive to electrons and so the final image is projected either on to a viewing screen coated with a material which fluoresces when irradiated with electrons, or on to a photographic plate if the image is to be recorded permanently. Since it projects the final image, the lens in an electron microscope which corresponds to the eyepiece of a light microscope is called the *projector lens*. The other lenses, *condenser* and *objective*, have the same names as in the light microscope and serve the same functions.

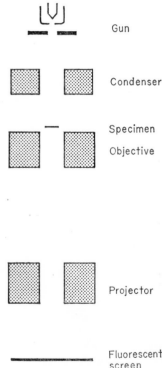

Gun

Condenser

Specimen

Objective

Projector

Fig. 1–5 Diagram showing the layout of a simple electron microscope.

Fluorescent screen

One extremely important property of an electron beam has not so far been mentioned. It is that it can be produced only in a vacuum. Electrons cannot travel more than a very short distance in air, since they are stopped by collision with gas molecules. The interior of an electron microscope must therefore be evacuated. The actual vacuum required is about 10^{-4} mm Hg, and this is obtained by means of vacuum pumps. As we shall see, the requirement that the microscope must work under vacuum imposes severe limitations on the kind of specimen that can be examined.

Simple microscopes of the kind depicted in Fig. 1–5 were constructed experimentally in the 1930's, and it was soon shown that their resolving power could be better than that of a light microscope. These early instruments were primitive in design and difficult to operate. Nevertheless, they contained all the essential features of the modern electron microscope, and the development of the latter has been mainly a question of incorporating large numbers of technical refinements, which have led to a progressive improvement in performance and ease of operation. The basic principles have remained unaltered.

1.7   Modern electron microscopes

Figure 1–6 shows the layout of a typical high-resolution electron microscope. It differs from the early instruments chiefly in that there are more lenses: instead of one condenser lens there are two, and instead of there being just two lenses to form the image of the specimen there are four, additional *intermediate lenses* being inserted between the objective and projector. One reason for having two condenser lenses is that they make it possible to produce a narrower illuminating beam, and thereby limit the area of the specimen which is irradiated by electrons. This may be important in order to prevent damage to the specimen. The intermediate lenses allow the magnification to be changed over a wide range. In practice the range in magnification of an electron microscope is usually between about × 1,000 at the bottom end of the scale (over-lapping with the light microscope) to about × 200,000 at the top. The very highest magnifications are not commonly used, however, since the maximum resolving power attainable with biological materials can usually be achieved with smaller enlargements.

The fluorescent screen on which the final image appears is viewed through thick glass windows in the chamber at the base of the microscope column. For critical focusing of the microscope, the image on this or another viewing screen can be examined through a low-power binocular microscope mounted outside the column. The screen can be tilted to permit this, and it can also be raised out of the way to allow photographs to be taken with the plate camera situated below. The whole microscope column, from the gun at the top to the plate camera below is about six feet high. It is usually fairly massively constructed, in order to ensure mechanical stability and prevent disturbance by vibration of the building, etc. The lens coils, through which moderately large currents are flowing, would tend to get hot and are therefore usually cooled by water circulating around them.

The various lenses and the gun all have to be accurately lined up on a common axis, and for this there are mechanical controls for moving them about. The specimen itself is moved on a special stage equipped with delicate controls operated by fine micrometer screws. With these the specimen can be moved accurately over small distances and placed in any desired position.

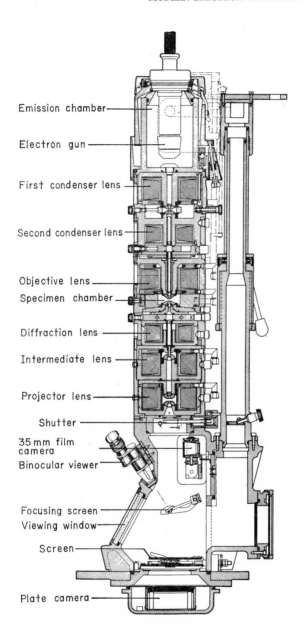

Fig. 1–6   Section of the column of a high-resolution electron microscope (Philips, EM 200).

As already explained, the interior of the microscope has to be evacuated, and this requires high-performance vacuum pumps, as well as a gauge to measure the pressure. The pumps are usually located in the desk on which the microscope column is mounted. In order to be able to change the specimen without breaking the vacuum in the column an air-lock is provided, through which the specimen can be introduced into the microscope with only a very small amount of air.

The other main components of the electron microscope are the electrical supplies, the most important of which are the high-voltage system for the electron gun and the circuits providing the lens currents. There are important reasons why these have to be fairly elaborate. All electron lenses suffer from defects. One of these is analogous to the fault found in many glass lenses and is called *chromatic aberration*: the focal length of the lens varies with the wavelength of the illumination. As already noted, the wavelength of an electron beam depends on the voltage used to accelerate the electrons, and if this voltage is not kept constant the wavelength will obviously vary. In practice the high-voltage supply has to be stabilized to about 1 in 100,000 in order to provide a sufficiently 'monochromatic' beam. The focal length of an electron lens also depends on the value of the current flowing in the coil, and this too must obviously be highly stabilized if the focal length is to remain constant. Again, the permissible variation is about 1 part in 100,000. Each lens is provided with a separate stabilizer. The precise value of each lens current can usually be controlled by means of resistances, and the lenses can also be switched off completely if necessary. The high-voltage supply of the gun and the filament-heater current can also be adjusted by electrical controls, so that an electron microscope usually has a fairly formidable control panel of knobs and switches. All the operations which in a light microscope are done mechanically, such as changing the objective or eyepiece to alter the magnification, or racking the body of the microscope up and down to bring the specimen into focus, are carried out in the electron microscope by adjusting electrical controls.

With its large column, numerous lenses, specimen stage, camera, pumping system and complicated electronic supplies the electron microscope has become a highly elaborate instrument, hardly recognizable as the offspring of the simple experimental microscopes made just over thirty years ago. Nevertheless, that is what it is. Beneath all the refinements and elaboration the basic principles remain unaltered, and from the point of view of understanding the role of the electron microscope in biological research it is these that matter.

It remains only to consider the resolving power of the modern electron microscope. As already noted, the wavelength of electrons is only a fraction of an Ångström unit, but the resolving power which this should theoretically make possible has not been achieved in practice. There are several reasons for this. One important one is that all electron lenses are highly imperfect compared with the glass lenses of the light microscope. Apart from their chromatic aberration, which has already been mentioned, they also suffer from spherical aberration, which means that their focusing action varies in passing from the optical axis outwards to the periphery. In the light microscope this defect can be largely overcome by combining lenses of different shape, which correct each other's faults, but this approach is not yet possible with electron lenses. The solution which has to be adopted (and which is also used to some extent in the light microscope), is to work with only the central portion of the lens, and this is done by building into it an aperture of small diameter. The restriction in the aperture of the lens cannot be carried too far, however, without deterioration of image quality arising from other sources, and in practice a compromise value has to be adopted. Spherical aberration in the lenses is one of the main obstacles preventing the attainment of the theoretical resolving power, but it is not the only one, and a number of other factors are involved. For example, as the resolving power is improved it becomes necessary to make a corresponding improvement in the mechanical stability of the microscope. Obviously, if we are hoping to achieve a resolving power of a few Ångström units the specimen must not move by that amount while it is under observation. Ensuring this can be a matter of considerable difficulty, since we are dealing here with distances comparable to those which separate atoms in molecules or crystals. The column and specimen stage have to be specially designed to eliminate vibration and movement, and precautions have to be taken to keep the specimen at a stable temperature, otherwise thermal expansion will cause it to move about.

The resolving power which is actually achieved depends largely on the quality of design and construction of the microscope, and also on the skill and experience of the operator. It also depends on the nature of the specimen being examined. The best results which have been obtained so far come from work on certain crystals, in which it has been possible to resolve spacings in the crystal lattices of just over 2 Å. With biological specimens the resolving power is not at present as good as this, and is more likely to be in the region of 10 Å. However, compared with the resolving power of the light microscope—0·2 μ or 2,000 Å—this represents an improvement of about 200 times, which is adequate for most purposes at present.

1.8   Other types of electron microscope

The type of microscope which has just been described is a conventional, high-resolution instrument of the kind which is most commonly used in biological laboratories. A number of smaller instruments, of lower resolving power (20–100 Å) are also available, and are used for work not requiring the best possible performance. These have fewer lenses and less elaborate electronics. In addition to these, however, there are various special types of microscope which make use of electron beams and electron lenses but are otherwise considerably different in their method of operation. A detailed description of these lies outside the scope of this booklet, but one example which may be mentioned is the scanning microscope, which is chiefly used for examining the surfaces of solid objects. In this instrument a fine beam of electrons is made to scan the specimen, just as an electron beam scans the screen in a television tube, and an image is built up sequentially. Depending on the chemical composition and topography of the surface of the specimen, electrons are reflected or absorbed to different extents in different regions. The reflected electrons can be collected and made to form an image on a fluorescent screen by suitable electronic devices. The resolving power of such microscopes is limited by the size of the electron beam used to scan the specimen, and is rarely better than 100 Å. For examination of surfaces, however, this is an extremely useful type of microscope, and an example of its application will be given in Chapter 4.

Viruses and Molecules 2

2.1 Introduction

The development of the light microscope as a tool of biological research which went on in the eighteenth and nineteenth centuries, depended not only on improvements in the microscope itself but also on the invention of increasingly sophisticated techniques for preparing material for examination. At the outset the early microscopists were largely restricted to examining small objects that scarcely required any preparation. They studied protozoa or other minute organisms from pond water, looked at insect mouthparts, examined pollen, feathers, hair and butterfly scales. All these could be placed under the microscope intact. To use the microscope to study the internal structure of larger organisms required greater ingenuity. One of the first methods to be developed, and which was used a great deal at one time, involved soaking small pieces of tissue in solutions which dissolved the material binding the cells together, causing them to fall apart. This process was called maceration. With favourable material it could yield useful information about the shapes of cells, and even some details of their internal organization. Another simple technique consisted of making smears or squashes of small pieces of tissue. Provided the cells separate or flatten sufficiently, a great deal of their internal structure can be made out in such preparations. Squashes are still commonly used for studying chromosomes. These methods, however, will work with only some kinds of material, and are of rather limited scope when it comes to elucidating in detail the three-dimensional structure of tissues and organs. This requires a technique which preserves the original shapes and mutual relationships of the cells, and yet permits these to be seen. The only way to do this is to cut the material into a series of thin sections, and methods for doing this were worked out in the nineteenth century. Concurrently with the development of all these techniques for producing specimens small enough and thin enough to be studied there was a great deal of research into methods for preserving cells and tissues, so that they would withstand subsequent processing, and for staining them, so as to make their structure more clearly visible.

Details of some of the methods used in light microscopy will be touched upon below, where they are relevant. What is important to note now is that the light microscope became a really useful tool for the biologist only when the problems of preparing material for examination had been satisfactorily worked out; for precisely the same thing happened when the electron microscope became available. The preparation of material in suitable form presented even greater problems than it did with the light microscope, and a great deal of effort and ingenuity had to be expended to devise satisfactory techniques. For reasons which will become clear later, it is important that at least the outlines of these techniques should be understood, and they will be discussed in some detail in the following pages, in which the applications of the electron microscope in biological research are described. The first objects which will be considered are viruses, which were among the earliest biological specimens to be examined in the electron microscope.

2.2   Preparative techniques for viruses

All viruses are exceedingly small, but some are smaller than others. A few of the larger ones can just be seen in the light microscope as minute dots, but their shape cannot be made out and nothing can be seen of their internal structure. Before the advent of the electron microscope the shapes and dimensions of viruses had to be deduced indirectly; for example, by

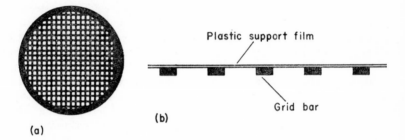

(a)

(b)

Plastic support film

Grid bar

Fig. 2–1   (a) A grid for supporting specimens (actual diameter 3 mm).
(b) part of a grid and plastic support film, in section.

determining how fine a filter they would pass through, or by studying their rate of sedimentation in a centrifuge. Such methods could give only very approximate results. Almost nothing was known about the internal structure of viruses. Before considering how the electron microscope has changed this situation we must describe the methods by which viruses are prepared for examination.

Specimens which are to be studied in the electron microscope have to be supported on something, and this obviously cannot be a glass slide such as is used in the light microscope, since this would be opaque to electrons. Instead of a slide use is made of a fine metal grid or mesh, 2–3 mm in diameter and usually made of copper, with holes about 0·1 mm across (Fig. 2–1). (Grids of this kind are usually made by electrolytic deposition on a suitably shaped cathode.) The grid is covered before use with a thin plastic film, commonly of nitrocellulose (celloidin), which forms the actual support for the specimen. Such films are prepared by dissolving the plastic in a volatile solvent and allowing a drop of the solution to spread on a water surface; the solvent evaporates, leaving behind a sheet of plastic only a few hundred Ångström units thick. This is thin enough to be transparent to electrons, yet strong enough to withstand irradiation.

Viruses are usually obtained from infected plants or animals by grinding up pieces of tissue. This has to be done vigorously, in order to break up the cells in which the viruses multiply, and it is usually done in a special piece of apparatus called a homogenizer. The technique is considered more fully in the next chapter (p. 42). Alternatively, it is sometimes possible to obtain viruses from the body fluids of infected organisms. However obtained, the suspension or fluid can then either be examined without further treatment, in which case the virus particles will occur mixed up with all sorts of cell debris and other material, or else steps can be taken to extract the viruses and free them from contamination. In outline, the latter process commonly involves the use of a centrifuge to spin out the heavier cell fragments, etc., leaving the virus particles in suspension. The latter can then be concentrated by centrifuging at higher speed, and they can be washed if necessary by resuspending them in a suitable saline solution and centrifuging again. Many technical variations are possible in this procedure, but the end result aimed at is always a pure preparation of virus particles, uncontaminated by foreign material. Small drops of the suspension of virus particles may then be placed on plastic-covered specimen grids, and allowed to dry. Often the suspension is sprayed on to the grids through an atomizer, which produces conveniently small droplets.

If such a preparation were examined in the electron microscope without further treatment the results would almost certainly be disappointing. The virus particles would probably be seen, but they would be indistinct, and scarcely stand out against the background. It would almost certainly not be possible to make out any internal structure in them. The reason for this is important, since it applies not only to viruses but to biological specimens in general. We therefore have to discuss now why the image of such virus particles should be so poor in contrast and lacking in detail, and to describe what must be done in order to improve it. To do this we must leave viruses for the moment and consider how an image is formed in the electron microscope.

To understand image formation it is simplest to revert to considering electrons as particles. An electron beam consists of a stream of such particles, accelerated down the microscope column by the high voltage of the electron gun. As already explained, electrons are readily stopped or diverted by collision with atoms or molecules, which is the reason for working in a vacuum and also for supporting the specimen on an extremely thin plastic film. The specimen itself is, of course, made of atoms and will therefore also act as an obstacle to the passage of electrons. In passing through the specimen some of the electrons in the beam will encounter atoms, and as a result of the collision will be diverted or scattered. If they are scattered through a sufficiently large angle they will be lost to the electron beam, since all the lenses of the microscope are fitted with narrow apertures (see

p. 13). Other electrons, however, will pass straight through the specimen and will end up on the viewing screen or photographic plate (Fig. 2–2). The extent to which electrons will be scattered by any particular part of the specimen depends on two things: the number of atoms per unit volume, and the size of the atoms in question. The more atoms there are in a given region, and the larger they are, the greater is the probability that an electron passing though will collide with one of them. The dark areas in the final image therefore correspond to regions in the specimen which scatter large numbers of electrons, thereby removing them from the beam, while the light areas correspond to less dense regions which allow electrons to pass freely. The distribution of densities in the final image is closely related to the distribution of matter in the specimen.

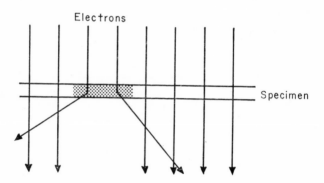

Fig. 2–2 Diagram to illustrate how electrons are scattered in passing through a dense region of the specimen (stippled), but not by less dense regions.

The size of atoms can conveniently be expressed in terms of the atomic number of the element in question. Elements with high atomic numbers are highly effective in scattering electrons, those with low ones are not. This fact explains why it would have been possible to make out very little detail in a preparation of viruses made simply by allowing the particles to dry down on a support film: viruses, like living organisms in general, are made up almost entirely of light elements, such as carbon, hydrogen, oxygen, nitrogen, phosphorus and sulphur.

In order to obtain satisfactory images of viruses or other biological specimens in the electron microscope it is necessary to treat them in some way so as to enhance their electron-scattering power. This is analogous to staining cells or tissues for examination in the light microscope. In 'staining' for the electron microscope we make use not of dyes, which selectively absorb light of certain wavelengths, but of certain salts of heavy metals, such as lead, uranium or tungsten, which have high atomic numbers. In the simplest kind of procedure the material to be examined is soaked in a solution of the stain (for example, uranyl acetate) and, if it combines with this sufficiently, its visibility in the electron microscope will be increased. The success of this kind of staining method depends on both the specimen and the stain. Proteins, however, which are present in all living organisms, quite readily combine with salts of heavy metals and it is therefore usually not difficult to achieve some degree of staining.

The simple and direct method of staining just described is used a good deal for cells and tissues, as will be explained in the next chapter, but it is not much used for viruses, since there are other, more effective techniques for making their shape and structure visible in the electron microscope. These also make use of the electron-scattering power of heavy metals, but do so in different ways. In the simplest method the stain is used not to impregnate the virus particles but to form a thick, opaque layer around them. The virus particles, being largely unstained, then stand out as light

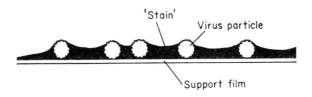

Fig. 2–3   Diagram to illustrate the technique of negative staining, in which the specimen is made to stand out against an electron-opaque (dark) background.

objects against a dark background. This is called 'negative staining' (Fig. 2–3). The method is simple in practice, since it is only necessary to mix the virus particles with the solution of stain and to spray the mixture on to grids covered with support films. Uranyl acetate is one of the substances commonly used as the 'stain' in this method, and so are sodium or potassium phosphotungstate. These substances dry out to form almost structureless solids; they surround the virus particles very closely and fill all the crevices and hollows in their surfaces, so that a great deal of minute structural detail may be revealed, as we shall see.

The second technique commonly used for studying viruses was developed earlier than the negative-staining method, but is somewhat more complicated. In this case heavy metals are again used as electron-scattering materials, but in the form of the metal rather than as salts. The technique is called 'shadowing' and is illustrated in Fig. 2–4. A preparation of virus particles is made on a grid in the usual way, without staining, and then a thin layer of metal is deposited over it. This is done by evaporation: a piece of metal wire is heated to incandescence, in a vacuum chamber, with the specimen placed nearby. If the specimen is placed not directly facing the metal source but at an angle to it, as in Fig. 2–4, the metal will not be deposited in the 'shadow' of objects (such as virus particles) standing up above the level of the support film. The shadowing technique does not reveal internal structure in viruses, but it can give fairly precise information about their shape and dimensions. This is particularly the case if care is taken to determine the angle of shadowing accurately, since the length of the shadow then enables the height of the specimen to be calculated.

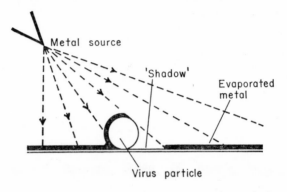

Fig. 2–4   Diagram illustrating the shadowing technique.

The two techniques—negative staining and shadowing—illustrate the important point that most biological specimens are of too low electron-scattering power to be studied in their original form in the electron microscope, and that the enhancement of contrast depends on the use of heavy metals. Further aspects of specimen preparation will be considered in the next chapter, when the structure of cells and tissues is considered. The remainder of this chapter will be devoted mainly to describing some of the facts which the electron microscope has revealed about virus structure.

2.3   Viruses

The electron microscope quickly yielded a large amount of new information when applied to viruses. At the outset many different sorts of viruses were examined and it was found that each had a characteristic shape and size. Many were approximately spherical, ranging in diameter from not much more than 100 Å to almost 0·5 μ in a few cases. Others were rod-shaped, and some (the viruses attacking bacteria) were more complicated in form. The examination and description of virus particles of different kinds was a necessary preliminary to more detailed experimental work on their chemical composition, and on the processes by which they are produced inside living cells. Research on viruses, just as on any other kind of organism, depends on their accurate description and characterization in the first place, since without this it may be impossible for the observations of one worker to be repeated by another. Taxonomy is still a fundamental part of biology, even at the level of viruses.

Apart from enabling the shapes and sizes of different kinds of viruses to be described, the electron microscope also provided an accurate method for counting virus particles. In quantitative experiments on virus growth it is usually necessary to be able to measure the concentration of virus particles, and in many cases this is not easy unless direct counting is possible. In one of the simpler methods for counting viruses the suspension containing them is mixed with a known concentration of polystyrene latex particles, which are small plastic spheres about the same size as virus particles, and the mixture is then sprayed on to grids with an atomizer. The contents of each droplet dry down separately and the ratio of virus to latex particles is estimated from counts made in the electron microscope. Since the concentration of latex particles is known, that of the virus can be calculated.

The electron microscope, then, has enabled us to examine and count virus particles. It has done more than this, however, for it has also been an indispensable tool in recent research which has extended our understanding of viruses down to the molecular level. The structure of some viruses can now be described to a considerable extent in terms of the different types of molecules of which they are made and the way these molecules are fitted together, and a good deal is also known about the manner in which new virus particles are formed inside cells. It is essential to note, however, that although the electron microscope has played an important part in this work, it has been only one among many methods of investigation. The detailed picture of virus organization which is now emerging owes just as much to biochemical and genetical studies, and to immunological work, as it does to observations with the electron microscope, and much of the information provided by the electron microscope would be difficult to interpret without the results obtained using other techniques. In the remainder of this section we shall confine ourselves mainly to a discussion of some of the principal features of virus structure, rather than their growth in cells, since it is here that the contribution of the electron microscope has been particularly important.

It is known from chemical analyses of purified preparations that most virus particles consist entirely of protein and nucleic acid. Except in a few special cases no other type of molecule is present. The first question we have to consider, therefore, concerns the way in which the protein and nucleic acid are arranged in the virus particles. Here it is possible to make an important generalization. In spite of their diversity in shape and size, all viruses prove to be constructed on an essentially similar plan: the protein forms an outer case or shell, and the nucleic acid is inside (Fig. 2–5). This is shown by various pieces of evidence. With some viruses it is possible to separate the protein from the nucleic acid simply by rapidly diluting the medium, which subjects the particles to 'osmotic shock'. After doing this it is found that the virus particles look empty in the electron microscope, instead of having dense contents. Chemical analysis shows that their nucleic acid has been liberated into the medium, while the empty particles consist of protein alone. It has also been found that the nucleic acid of most viruses is protected from attack by enzymes as long as the particles are intact, but becomes susceptible if the particles are damaged. There is evidence from various sources that, of the two components, it is the nucleic acid which is essential for the reproduction of the virus, since it is infective by itself, whereas the protein is not. The protein can therefore be regarded as forming a protective shell around the nucleic acid, preventing it from damage by the environment.

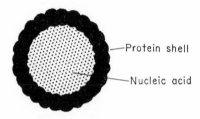

Fig. 2–5   Diagram illustrating schematically the basic structure of a virus.

Plate 1c (facing p. 28) shows an electronmicrograph of one of the spherical viruses, human wart virus, prepared by the negative-staining technique. As already noted, this technique shows up surface structures. It will be seen that the particles do not have smooth surfaces but are apparently made up of small globules, or subunits. These are outlined by the stain, which has penetrated into the grooves between them. We know that in this micrograph we are seeing the outer surface of the virus particles, and are therefore looking at the protein coat. What, then, do the globules represent? The answer is that they must be either single protein molecules or small aggregates of them. Several lines of evidence point to this conclusion. First, it agrees with what is known about the dimensions of protein molecules. Although proteins are long-chain molecules, made up of large numbers of amino acids, they do not usually occur in completely extended form. The polypeptide chain is commonly folded up in an intricate manner, giving rise to a molecule which is approximately globular in overall shape. Such globular molecules are usually a few tens of Ångström units in diameter. Second, in the case of one virus (tobacco mosaic virus, Plate 1a and b) it has been possible to make a very detailed study of the protein coat, using a variety of methods in addition to electron microscopy, and it has been shown that the molecules which form it are ellipsoidal in shape, about 60 Å long and 25 Å in diameter, and that these are regularly and closely packed in a layer one molecule thick. Such detailed information about the size and shape and arrangement of the protein is not available for most other viruses, but the appearance of more or less globular subunits in negatively stained preparations of many kinds of virus particles suggests that the basic arrangement is probably very similar.

For the time being, since it is not certain in most cases whether the globules we see in electronmicrographs of virus particles are single protein molecules (as in tobacco mosaic virus) or small aggregates of them, the non-committal term 'subunit' is used. It will be noted in Plate 1c that the subunits appear to be regularly arranged; that is, they are not scattered at random but are evenly spaced over the surface of the particles. This is found to be the case with all viruses of which sufficiently good electron-micrographs have been obtained. Careful analysis of the micrographs shows that the pattern in which the subunits forming the protein coat are arranged is always highly constant in any given type of virus, and is of a crystalline degree of regularity. In other words, the subunits are fitted to-gether almost as precisely as are the atoms in a crystal of salt. The scale of the two kinds of crystal is, of course, very different; each of the virus sub-units consists of at least one protein molecule, itself composed of thousands of atoms. Nevertheless, the regularity of spacing which is achieved in the two cases is comparable. In order for the subunits to maintain constant spacings and mutual relationships they must be held together in some way, and this appears to be effected mainly by weak linkages such as hydrogen bonds. It is postulated that each type of subunit must have a characteristic pattern of chemical groups on it surface, and that these are able to form linkages with suitably located groups on other subunits.

For reasons which need not be discussed here, there is only a small num-ber of geometric arrangments in which it is possible to fit together subunits to form cylinders or spheres such as are commonly found in virus shells. The detailed molecular architecture of the protein component of viruses is restricted to a few standard patterns.

The next question to be asked, and which the electron microscope helps to answer, concerns the way in which the protein coat is put together when a new virus particle is made. As noted above, the protein molecules are regularly packed and joined by weak bonds. We now want to know how they come to take up their characteristic arrangement. Experiments on tobacco mosaic virus provide an answer to this question. If this virus is exposed to alkaline solutions (pH 10), the protein and nucleic acid separate from each other and the virus particles fall apart. The protein can then be separated in pure form. If a solution of this is made neutral, the protein aggregates spontaneously, forming cylindrical rods of the same size as the original virus particles. This can be demonstrated simply by examining sam-ples of the preparation in the electron microscope. If the same experiment is done but with the nucleic acid present then this too is incorporated and complete, infective virus particles are formed. The nucleic acid, however, is not essential for the assembly of the protein. This discovery, that protein molecules will spontaneously assemble if the environmental conditions are right, is extremely important. It means that we do not have to postulate any additional agency to fit the molecules together. No template or mould, for example is necessary to give the virus particles their characteristic shape. Just as sodium and chloride ions, if brought together under appropriate conditions, will automatically build salt crystals, so will virus proteins spontaneously build virus shells.

It seems likely that most viruses are to a very large extent 'self-assembling' in this way, and that this is the normal manner in which they are produced inside cells. Essentially, virus multiplication consists of synthesis of their component molecules (protein and nucleic acid) inside the host cell, followed by spontaneous assembly of the molecules into complete particles.

These findings about the construction and assembly of viruses are significant not simply because they help us to understand the organization of an important class of disease-causing agents, but because the essential ideas can be extended beyond the viruses to many different kinds of structures found in cells. We may take as one simple example the flagella of some bacteria. These are extremely delicate fibres, about 150 Å in diameter and therefore visible only in the electron microscope. They occur either singly or in bundles, and they are responsible for the movements of bacteria, just as are the cilia or flagella of some plant and animal cells. Plate 1d shows an electronmicrograph of bacterial flagella, prepared by the negative-staining technique. The flagella have a very regular beaded structure, each being made up of a number of rows of subunits which resemble those seen in virus particles. They correspond to single protein molecules or to small groups of them. It is known from chemical analyses of isolated flagella that they are made up almost wholly of one kind of protein. This can be isolated in pure form and, given the right conditions, it will spontaneously aggregate to re-form flagella. The parallel between these flagella and tobacco mosaic virus, both in the regularity of structure at the molecular level and in their self-assembling properties, is very close.

Similar studies have been made on other protein components of cells; for example, the fine filaments of striated muscle (see p. 47). The full significance of this work will become clearer in the next chapter, in which we shall consider what the electron microscope reveals about the structure of plant and animal cells. For the time being it is sufficient to recognize that with the aid of the electron microscope it has been possible to describe the structure of living organisms in some detail in terms of the molecules of which they are made up and the way those molecules are fitted together.

2.4   Molecules

It will be clear from the preceding section that some large molecules are directly visible in the electron microscope. They can be studied either by the negative-staining method, or by shadowing. From the biologist's point of view the most interesting of these large molecules are the proteins and nucleic acids. A considerable range of proteins has been examined, and these often turn out to be more or less globular, like those making up the protein coats of viruses. Many enzymes are like this. In these cases, of course, the polypeptide chain which forms the protein is tightly folded up. There are other proteins which are less tightly folded and which may appear long and thin in the electron microscope. The protein which makes up the collagen fibres of connective tissue is of this type. Nucleic acid molecules have also been studied, and especially deoxyribonucleic acid (DNA), which is found in the chromosomes and is the hereditary material of most organisms. This appears in the electron microscope as a long thin thread, rather like a piece of string, but only about 20 Å in diameter.

While the overall shape of these large molecules can be made out, the electron microscope reveals little about their detailed form. This is partly because the resolving power of the electron microscope is too low, partly because of the technical difficulties of preparing specimens in a suitable form. Perhaps in future some of these obstacles will be overcome; in the meantime, however, the study of molecular structure will continue to be carried out chiefly by indirect methods.

Cells and Tissues

3.1 Techniques of preparation

As already explained, small objects like viruses or molecules can be examined directly in the electron microscope, without elaborate preparation. All that is necessary is to use some method for making detail more clearly visible, which in practice means either using negative staining or shadowing. When we come to study the cells and tissues of plants and animals, however, the situation is very different, and it is necessary to use rather elaborate procedures for preparing material for examination. There are two main reasons why this is so. First, living tissues, unlike viruses, contain at least 70 per cent water. The electron microscope works at high vacuum (see p. 9), and if living cells were placed in it they would immediately be dehydrated by evaporation. They would collapse, and their intricate structure (to be described later in this chapter) would be largely destroyed. Second, even if this were not the case, it would still be impossible to make out much detail in whole cells. This is because they are too thick. A virus particle is not more than a few hundred Ångström units thick, whereas a cell is usually between 10 and 100 μ—that is, at least a thousand times thicker. The reader should be able to deduce the disadvantages of this greater thickness by considering what was said in the previous chapter about the mechanism of image formation in the electron microscope. It will be remembered that an object becomes visible if it scatters electrons to a different extent from its surroundings. In the case of a thick object, however, such as a whole cell, most of the electrons passing through it will be scattered, since there is a high probability that they will encounter atoms on their way through. The image therefore becomes uniformly dark, and contains little detail. The only way round this problem is to examine not whole cells but pieces of them, and in practice this usually means that they have to be sectioned.

The technique of cutting plants and animals up into thin slices, or sections, in order to study their internal structure, has been used for a long time by light microscopists, and the reader will probably be familiar with the sort of sections used in histology. The sections used in the electron microscope are essentially similar, but the whole procedure used in preparing them has to be greatly refined. For the light microscope the sections are usually between 5 and 10 μ thick. Occasionally they may be as thin as 2 μ, though with the techniques usually employed it is difficult to cut sections as thin as that. For the electron microscope, however, the sections must not be more than 0·1 μ (= 1,000 Å) thick, and ideally they should be nearer to 500 Å. This extreme thinness is, of course, necessary in order to prevent excessive electron scattering and consequent loss of detail. In simple terms, thicker sections would be opaque to the electron beam.

To cut sections which are only about two millionths of an inch thick is not a simple matter, but the problem was essentially solved in the early 1950's, soon after high-resolution electron microscopes became commercially available. It is rather important that the reader should understand clearly the methods used, since otherwise it is difficult to appreciate either the potentialities or the limitations of the electron microscope as a tool for studying the structure of cells and tissues. The procedure will therefore be described here in some detail.

Living cells are fragile, watery objects, easily damaged by unfavourable conditions. The first step in preparing sections of them is to stabilize their structure, so that it will stand up to the rigours of subsequent treatment. This is done by treating them with a suitable reagent, called a *fixative*. Numerous chemicals have been tried as fixatives and a considerable range is still in use for work at the light-microscope level. For electron microscopy, however, only two fixatives are used at all commonly: osmium tetroxide (OsO_4), and certain aldehydes. In general terms, fixation of a cell or tissue depends on the formation of chemical bonds or cross-linkages between some of the molecules which make it up. Any cell, of course, contains an immense number and variety of molecules, and it is chiefly the proteins and lipids which are involved in fixation. The type of reaction which occurs can be illustrated by considering how aldehydes react with amino ($-NH_2$) groups, which are found on the side chains of protein molecules:

$$\underset{\substack{\text{lysine}\\\text{residue}}}{} \overset{\displaystyle \overset{|}{\underset{|}{NH}}}{HC}(CH_2)_4NH_2 + \underset{\substack{\text{formal-}\\\text{dehyde}}}{HCHO} \longrightarrow \overset{\displaystyle \overset{|}{\underset{|}{NH}}}{HC}(CH_2)_4N{=}CH_2 + H_2O.$$

$$\underset{\substack{\text{protein}\\\text{chain}}}{\overset{|}{\underset{|}{C{=}O}}} \qquad\qquad \overset{|}{\underset{|}{C{=}O}}$$

If we use a dialdehyde as fixative (for example, glutaraldehyde, $O{=}CH-CH_2-CH_2-CH_2-CH{=}O$) then this reaction can take place with amino groups on two different protein molecules, so that the fixative forms a cross-link between them. Similar reactions can take place with other groups. Since almost any protein will have many sites at which aldehydes can react multiple links can be formed, involving many molecules, and the proteins of the cell are converted into a stable, cross-linked network. Osmium tetroxide reacts, among other things, with double bonds in unsaturated lipids and also in proteins, and it can cross-link these molecules according to the following type of equation:

$$\begin{array}{c} | \\ CH \\ \| \\ CH \\ | \end{array} \quad \begin{array}{c} O \\ \diagdown\diagdown \\ \quad OsO_2 \\ \diagup\diagup \\ O \end{array} \longrightarrow \quad \begin{array}{c} | \\ CH-O \\ \diagdown \\ \quad\quad OsO_2 \\ \diagup \\ CH-O \\ | \end{array}$$

$$\begin{array}{c} | \\ CH-O \\ \diagdown \\ \quad\quad OsO_2 + \\ \diagup \\ CH-O \\ | \end{array} \quad \begin{array}{c} | \\ HO-CH \\ | \\ HO-CH \\ | \end{array} \longrightarrow \quad \begin{array}{c} | \\ CH-O \\ \diagdown \\ \quad\quad Os \\ \diagup \\ CH-O \\ | \end{array} \begin{array}{c} | \\ O-CH \\ \diagup \\ \| \\ \diagdown \\ O \quad O-CH \\ | \end{array} \quad + H_2O$$

The full details of the chemistry of fixation are not by any means under-stood, and this is an important gap in our present knowledge. In general, however, it is clear that fixation results in the stabilization of structure by the formation of chemical cross-links, chiefly involving proteins and lipids. Other constituents of the cell, which do not react with fixatives, will not necessarily be preserved and may be washed out during subsequent pro-cessing. This is particularly liable to happen with small, soluble molecules. On the whole, only the macromolecular skeleton of the cell is likely to be preserved.

The delicate organization of the cell is easily damaged or destroyed by adverse conditions and it is essential that fixation should be carried out quickly. Animals are therefore anaesthetized and dissected rapidly, and small pieces of the appropriate tissue are cut out and immersed in fixative

---

Plate 1. (a) Tobacco mosaic virus, negatively stained. The particles appear light against a dark background of stain. Note that they are uniform in width. The variation in length results from breakage. The particles have a hollow core, which is filled with stain and appears as a central dark line. Magnification × 140,000.

(b) A single particle of tobacco mosaic virus, in which the individual pro-tein subunits can just be made out (arrows). Magnification × 400,000.

(c) Human wart virus, negatively stained. Note the regularity in size of the spherical particles. The subunits forming the protein shell are clearly visible. Magnification × 150,000.

(d) Parts of three bacterial flagella, negatively stained, at very high magni-fication, showing the protein subunits (white globules) of which they are composed. Magnification × 650,000.

(a–c kindly supplied by DR. J. FINCH; d kindly supplied by DR. J. LOWY from J. LOWY and J. HANSON, 1965, *J. molec. Biol.*, **11**, 293–313, Plate III.)

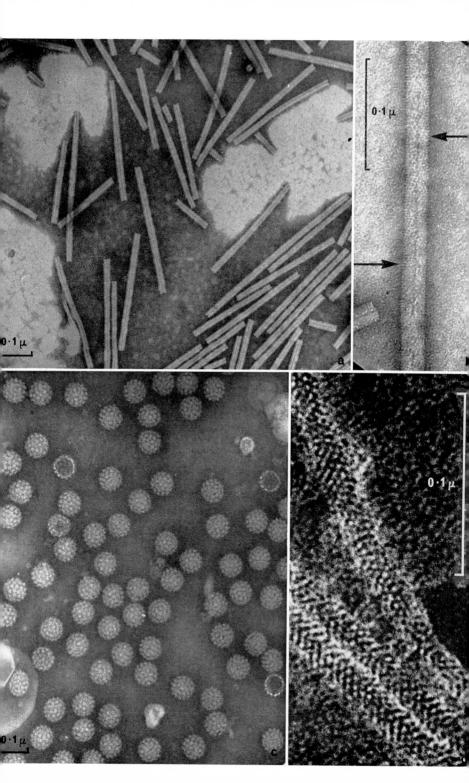

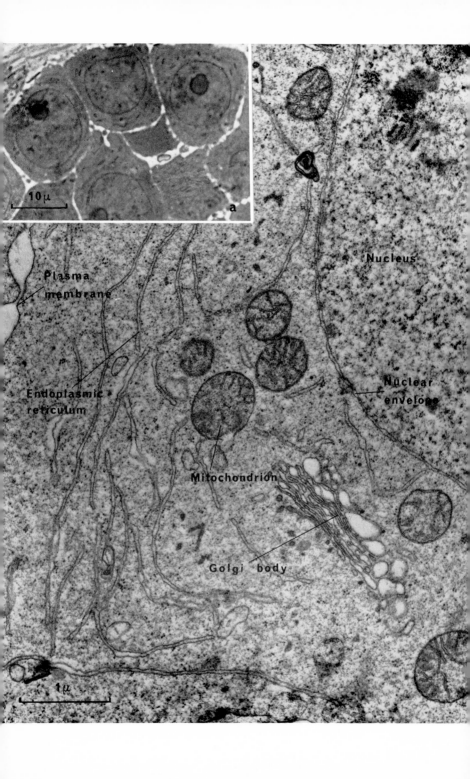

10μ

a

Plasma
membrane

Nucleus

Endoplasmic
reticulum

Nuclear
envelope

Mitochondrion

Golgi body

1μ

as quickly as possible. The samples are usually not more than one or two millimetres in thickness, so that the fixative can penetrate rapidly. Fixation usually lasts for about an hour or two, after which the excess fixative is washed out.

The material is still not by any means in a suitable condition for sectioning after fixation. It is too soft and has too little mechanical strength. If we tried to cut thin sections they would simply fall apart. The next step, therefore, is to embed the tissue in some material which will give it support during sectioning. The substance which has long been used for this purpose is paraffin wax, and most of the histological sections studied in the light microscope are prepared from paraffin-embedded material. Paraffin wax is not, of course, miscible with water, so after fixation the tissue has to be dehydrated. This is done by soaking it in ethanol/water mixtures of progressively increasing concentrations, starting with 50 or 70 per cent and ending up with pure ethanol. The tissue can then be transferred to molten wax, after first soaking it in a liquid such as benzene which is miscible with both ethanol and the wax. The specimen remains in the hot wax until completely infiltrated, and is then allowed to cool. The result is a small block of wax with the tissue inside, and this can be sectioned.

Unfortunately paraffin wax, although simple to use, cannot easily be cut into sections less than about 2 μ thick, and this precludes its use in the preparation of sections for the electron microscope. For the latter it is necessary to work with harder materials, and in practice various plastics are used. The principles involved are just the same as with paraffin, and the specimen has to be dehydrated as before, but then instead of infiltrating with the molten embedding material the unpolymerized form of the plastic is used. When this has penetrated the specimen completely the plastic is polymerized, usually by application of heat. The plastics chiefly used are various methacrylates (some of which are also the basis of Perspex) and certain epoxy resins (for example, Araldite), which are complex substances originally developed for use as adhesives. Both of these are strong materials, yet soft enough to permit sectioning. They also polymerize without substantial change in volume, so that distortion of the specimen is kept to a minimum.

Section cutting is the most delicate operation involved in the whole process of preparing cells and tissues for examination. Once again, the methods

---

Plate 2. (a) Photomicrograph of a section through an insect testis, showing a number of spermatocytes. The picture was taken using an oil-immersion objective. Apart from the nuclei (with prominent nucleoli), little detail of the structure of the cells can be made out here. Magnification × 2,000.

(**b**) Electronmicrograph of a section of part of one of the cells shown in (**a**). For detailed explanation, see p. 32. Magnification × 32,000.

(**b** kindly supplied by DR. A. M. MULLINGER.)

used are adaptations of techniques originally developed for cutting sections
for the light microscope (Fig. 3–1). The apparatus used is called a *microtome*. This is essentially like a bacon-slicer, except that the knife is kept
stationary and the specimen is moved over it, being advanced by a fixed
amount at each cutting stroke. The knife used is commonly a glass edge,
prepared by carefully breaking up plate glass. Polished diamond edges are
also used, and have a somewhat longer life than glass. The specimen
advance is usually brought about by a very fine screw-drive operated by a
ratchet mechanism. It need hardly be said that the whole apparatus has to
be constructed to a high degree of precision in order for it to produce
consistently sections of the required degree of thinness. Care has to be
taken to avoid draughts or local heating from the operator's hands, etc.,
which would cause thermal expansion or contraction of the microtome and
upset its performance. The maximum area of the sections which can be cut
is quite small. Usually they are not more than half a millimetre across, and
often they are less than this. The tip of the plastic block containing the
specimen therefore has to be carefully trimmed down to the appropriate
size, and this is done with a razor blade while watching through a low-
power dissecting microscope (Fig. 3–1). The sections, apart from being
very small, are also extremely fragile, and would be almost impossible to
handle by themselves. This problem is overcome by fitting a small con-
tainer to one side of the knife and filling it with water, the level of which is
adjusted so that the meniscus exactly meets the knife edge. As the sections
are cut they float on the water surface (Fig. 3–1 d), and can be picked up
on grids of the kind already described. With practice it is possible to cut
long series of sections which remain adhering to each other and can be
picked up as a continuous ribbon. The whole process is observed under a
dissecting microscope, so that the sections can be examined as they are
cut.

Only one further operation now remains before the sections are examined
in the electron microscope. This is to 'stain' them. As already noted (p. 18),
cells and tissues have low intrinsic electron-scattering power, so that in
unstained sections little detail of their structure can be made out in the
electron microscope. It is essential to increase their electron-scattering
power artificially and, as with viruses, use is made of heavy metals to do this.
Instead of using them for negative staining or shadowing, however, these
substances are applied directly to the sections. They are taken up to different extents by different parts of the cells and tissues: uranyl salts, for
example, chiefly combine with proteins and nucleic acids, whereas lead
salts are taken up more strongly by lipid-containing membranes. The embedding plastic remains in position in the sections and is not removed.
Fortunately it is sufficiently permeable to water to allow staining to occur,
and it does not take up the stains itself.

Having described the technique of preparing thin sections, we can now
consider what we see in cells and tissues subjected to this process.

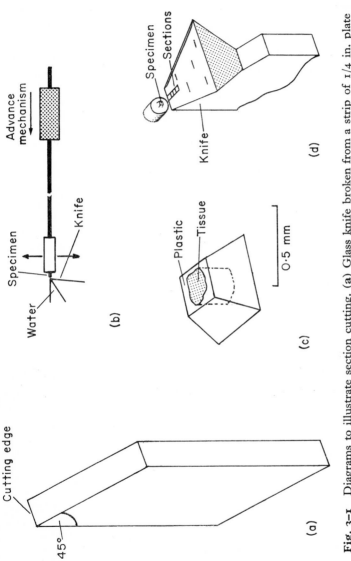

Fig. 3-1   Diagrams to illustrate section cutting. (a) Glass knife broken from a strip of 1/4 in. plate glass. (b) Basic layout of a microtome. (c) Plastic-embedded specimen, trimmed down and ready for sectioning. (d) A ribbon of sections floating at the knife edge on the water surface, ready to be picked up on a grid.

3.2   An outline of cell fine structure

The fine structure of cells is now a large topic, and to deal with it at all adequately would require far more space than is available here. In particular, it is impossible to include all the illustrations that would be necessary. Fortunately, there are now several published collections of electronmicrographs which display the features of cell fine structure in detail (see, for example, FAWCETT, 1966; HURRY, 1965; and PORTER and BONNEVILLE, 1964), and the reader needing a comprehensive account of the subject is advised to consult one or other of these books. All that it is intended to do here is to illustrate the sort of results that the electron microscope yields.

Plate 2a (p. 29) is a photograph of a section of some cells (insect spermatocytes) as seen with the light microscope. The section is about 2 μ thick and was photographed using an oil-immersion objective lens—that is, at maximum magnification of the light microscope. Plate 2b shows a thin section of part of one of these cells seen in the electron microscope. The two sections were cut from the same block. There is no difficulty in recognizing the main features of the cell in the electronmicrograph: the *nucleus* is easily identified by its position and shape, and at the left the edge of the cytoplasm is seen, bounded externally by a thin dark line, which is the *plasma membrane*. Provided we realize that in the electronmicrograph we are looking at a very thin, almost two-dimensional section of the cell, the transition from the customary light-microscope image to the less familiar electron-microscope picture is not difficult to make.

The chief difference between the light- and electron-microscope images lies in the much greater amount of detail visible in the latter, especially in the cytoplasm. Here, instead of the rather vague blobs and granules seen in the photomicrograph, there are sharply defined objects, some of them with intricate internal structure. The increase in the amount of detail visible arises, of course, from the greater resolving power of the electron microscope, and comparison of the two pictures provides a clear visual demonstration of the relative capabilities of the two instruments. The electronmicrograph shows some of the basic cytoplasmic components found in almost all cells, and these will now be briefly described.

The most prominent objects in Plate 2b are granules called *mitochondria*. These are approximately circular in outline, and therefore probably more or less spherical in three dimensions. Mitochondria are often of this shape, but in other cells they commonly take the form of short rods or filaments (Plate 3a). They are usually less than 1 μ in diameter but may be several microns long, so that they are about the size of bacteria. Their outer surface is formed by two parallel membranes, each about 70 Å thick. Inside there are other membranes, also arranged in pairs. In the mitochondria shown in Plate 2b these internal membranes are not very numerous, but in other cells they may be very abundant and almost fill the inside of the mitochondria (see Plate 3d, p. 44). Each pair of membranes has the form of a plate or shelf and is called a *crista*. The membranes forming the cristae are continuous at some points with the inner of the two bounding membranes (Fig. 3–2).

Mitochondria have been known since the late nineteenth century from studies with the light microscope, but with this they appeared simply as minute, structureless rods or spheres which could be demonstrated after fixing and staining cells by certain special techniques. The elaborate internal structure revealed in them by the electron microscope was an almost completely unexpected finding.

Plate 2b also shows another type of cytoplasmic organelle, namely a *Golgi body* (also called the Golgi apparatus). This appears in sections to consist mainly of a series of parallel dark lines, which are the cut edges of membranes. At first sight, therefore, the Golgi body consists basically of a

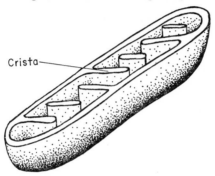

Crista

Fig. 3–2   Schematic reconstruction of a mitochondrion, cut open to show the arrangement of the cristae inside.

stack of membranes. Closer examination shows, however, that at their edges these membranes are joined together in pairs, each pair enclosing a flat space, so that the structure is more accurately interpreted as consisting of a pile of flattened sacs (this is shown diagrammatically in Fig. 3–3). In three dimensions the Golgi body as a whole may have the form of a flat or somewhat curved plate, or it may be irregular in shape. It will be noted that some of the flattened sacs have swollen margins, and that there are also many rounded vesicles of various sizes lying free in the cytoplasm around the Golgi body. It has been known for a long time that Golgi bodies are concerned with the formation of secretory products. In cells of the pancreas, for example, which form the proteolytic and other enzymes of the pancreatic juice, study of fixed and stained sections with the light microscope showed that the enzyme-containing granules which are ultimately liberated from the cell first make their appearance in the region of the Golgi body. This has been confirmed with the electron microscope, and there is now good reason to suppose that the secretory product is accumulated in the flat sacs, which swell up in consequence and transform into the round

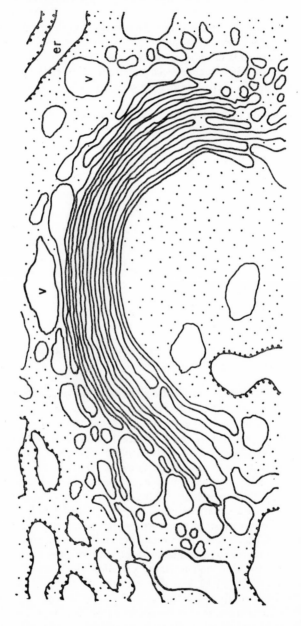

Fig. 3-3 Diagram showing a Golgi body as it appears in section. The flat sacs appear to be inflating to give rise to the vesicles (v) in the adjacent cytoplasm. Pieces of rough-surfaced endoplasmic reticulum (er), studded with ribosomes, are shown at left and right.

vesicles already noted. This may happen either on a large scale, in which case an entire sac may swell up, or it may be merely the margins of the sacs which swell and are pinched off. The vesicles thus produced probably themselves fuse together to form the large secretory granules which are ultimately discharged from the cell. The number of Golgi bodies in a cell varies a good deal and depends in part on how active it is in secretion. Quite often there is only one, but in some cells there may be over a hundred. Golgi bodies often, though not always, lie near the nucleus.

In addition to Golgi bodies there are several other types of membranous cytoplasmic structure. In Plate 2b these appear in the form of large single flattened sacs extending for a considerable distance through the cytoplasm. Membranes of this type occur very commonly in cells and vary considerably in their form and arrangement. In some cases they occur sparsely and widely separated from each other, but in others they may be abundant and closely packed, filling much of the cell (Fig. 3–4). Whatever their overall form, such membranes never occur as single sheets with free edges but always form the walls of sacs or tubules. In other words, they enclose spaces which are cut off from the general cytoplasm. Cytoplasmic membranes of this kind are frequently studded on their outer surfaces with small dense granules, 150–200 Å in diameter (Figs. 3–3 and 3–4). It has been shown that these granules consist of ribonucleoprotein, and they are usually called *ribosomes*. These may also occur free in the general cytoplasm.

The term *endoplasmic reticulum* (er) is commonly applied to all the cytoplasmic membranes, including both those with and those without attached ribosomes. Golgi bodies, which are also made of membranes, are sometimes included in this category. The term is something of a misnomer and was coined on the basis of early electron-microscope observations, which seemed to show that the cytoplasmic membranes were restricted to the inner (endoplasmic) region of the cell, and that they were all interconnected to form a network or reticulum. Neither observation was correct, but the term has persisted in spite of this and is now widely used. Those membranes with ribosomes on them are termed the 'rough (or granular) er', in contrast to the membranes forming the 'smooth er', which lack them. The functions of some of these different sorts of membranes will be considered below.

Mitochondria, Golgi bodies and the various types of membrane grouped together as the endoplasmic reticulum are the commonest and most characteristic cytoplasmic components and are found in plant and animal cells and in protozoa. Their exact form varies a good deal, as does their relative abundance, but they are always recognizable as the same types of structure. Any given type of cell—liver cell, neurone, lymphocyte, etc.—has a characteristic type of mitochondrion, arrangement of membranes and so on. There are some cells in which these cytoplasmic structures may be greatly reduced in number or even entirely absent. The red blood cells of mammals are the most extreme example: these are little more than bags of

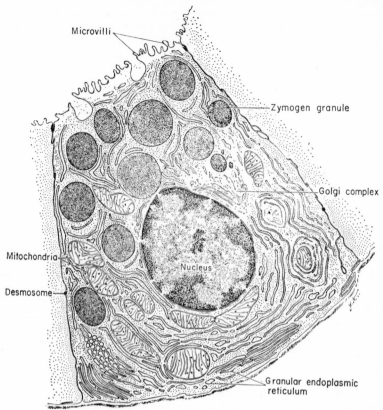

Fig. 3-4 Drawing of a secretory cell from the mammalian stomach, as seen in the electron microscope. The zymogen granules are enzyme-containing bodies which will ultimately be discharged from the cell. Much of the cytoplasm is filled with sheets and tubules of rough endoplasmic reticulum. Note the Golgi body. Desmosomes are structures concerned with the adhesion of cells to each other. The apical border of the cell, at the top of the picture, is drawn out into fine projections or microvilli. (Reprinted by permission of the authors and Rockefeller University Press from s. ITO and R. J. WINCHESTER. 1963, *J. cell Biol.*, **16**, 541–577, Fig. 18.)

haemoglobin, surrounded by a plasma membrane. Such absence of organelles is, however, a highly specialized condition, and any normally functioning cell may be expected to contain a basic set of organelles.

In addition to the structures so far described there are numerous other types of cytoplasmic component. For details of these the reader must refer to one of the books on cell fine structure already referred to, but the chloroplasts of green plants may be mentioned here. These, of course, are readily visible in the light microscope. The electron microscope reveals the details of their internal structure, which is seen to consist chiefly of an array of parallel membranes (Plate 3e). In algae the membranes are more or less equally spaced throughout the chloroplast, but in higher plants they come together in some regions to form tight stacks. Such regions constitute the *grana*, visible in the light microscope. Chloroplasts are enclosed externally by a continuous double membrane, like that of a mitochondrion.

The final type of cytoplasmic organelle for which there is space for description here are cilia and flagella. Plate 3b is an electronmicrograph of a flagellum from a protozoon. It is cut in transverse section, and it will be seen that it is approximately circular in section, and about 0·25 μ in diameter. The surface is formed by a membrane, which is continuous with the plasma membrane covering the rest of the cell. Below this there is a ring of nine structures, looking like figures-of-eight, and in the centre there is a pair of circles. Since this is a transverse section we can interpret all these as sections of longitudinally running, hollow tubules. The nine outer ones are double structures, each consisting of two tubules, while the central pair are single. This arrangement of a cylinder of nine outer tubules and two central ones is found universally in cilia and flagella (except bacterial flagella, see p. 24). Plate 3c shows cilia of the gills of a freshwater mussel and it will be seen that there is a remarkable similarity in their structure and dimensions to those of the protozoon shown in Plate 3b. It need hardly be pointed out that this is an unexpectedly complicated structure to be found in an organelle which can only just be made out with the light microscope.

It will be apparent even from this brief sketch that the electron microscope has revealed a surprisingly intricate structural organization in the cytoplasm, and this is one of the outstanding discoveries which the electron microscope has made possible. Previously, the existence of cytoplasmic membranes was largely unsuspected and little was known about the fine structure of any type of organelle. The cytoplasm, in fact, until quite recently, was regarded as a 'living substance' of rather mysterious, supposedly 'colloidal' composition. Now 'cytoplasm' is merely the name of a region of the cell (all of it except the nucleus), which is known to have an elaborate and precisely organized structure. The significance of this complex structure, from the point of view of the functioning of the cell, will be discussed later (section 3.4), but first we must complete this outline of cell structure by considering the *nucleus* as it appears in the electron microscope.

Compared with the wealth of structure which has been demonstrated in the cytoplasm, the electron microscope has not yet revealed a great deal in the nucleus of most cells. The most constant and best defined feature is the *nuclear envelope*, which forms the boundary between nucleus and cytoplasm (Plate 2b). This invariably consists of two membranes, each about 70 Å thick, running parallel to each other and separated by a rather wide gap (Fig. 3–5). In places there are circular 'pores', several hundred

Ångström units in diameter, which are probably regions through which materials may pass between nucleus and cytoplasm, though they are not completely open passages. The contents of the nucleus usually look rather uninteresting. Except in dividing cells the chromosomes are usually in a swollen, dispersed state and are often visible, if at all, only as regions of slightly greater density than the material between them (Plate 2b). Sometimes they are more readily recognizable, and particularly so in dividing cells in which the chromosomes are condensed. Even in these, however, little internal structure is visible. Fine fibrils, approximately 100 Å in diameter, appear to be the main structural component, and these are thought to represent fibrils of deoxyribonucleic acid (DNA) combined with

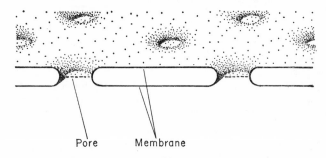

Pore   Membrane

Fig. 3–5   Diagram showing the nuclear envelope, formed from two membranes, with pores. Compare with Plate 2b.

protein. No overall pattern in the arrangement of these fibrils is visible, however, and the electron microscope has so far contributed little to our understanding of the organization of chromosomes. One reason for this is that chromosomes probably consist chiefly of very long, tightly packed and intricately arranged fibrils of deoxyribonucleoprotein, studying sections of which is rather like looking at sections through a skein of wool: we see lots of short lengths and cut ends, but the overall pattern escapes us. It seems unlikely that the study of sectioned cells in the electron microscope will tell us much about the organization of chromosomes.

3.3   The problem of artefacts

At this point, having given some account of what can be seen in cells with the electron microscope, it is necessary to break off and consider a fundamental, and rather difficult, question which can be formulated as follows. When we look at electronmicrographs of sections of cells we see all manner of complicated structures: these cannot be seen by any other method, because they are too small, and, for reasons which have already been explained (p. 26), they can be seen only in cells which have been subjected to an elaborate preparative procedure. This being the case, what reason have we for supposing that the structures we see bear any resemblance to the structures that existed in the living cell? How can we tell that the elaborate organization is not the product of our preparative technique? In short, how do we know that we are not producing *artefacts*? At the light-microscope level this problem does not arise so acutely, because it is usually possible to check observations on fixed and stained material by looking at living cells, for example by phase-contrast microscopy. In the case of electron microscopy, however, no such direct check is possible as yet. Recalling the violence of the treatments to which cells are subjected before they end up in the electron microscope—killing and fixing, the complete removal of water with organic dehydrating agents, impregnation with plastics, sectioning and staining—it would at first sight seem a matter for surprise if the end results were *not* artefacts, rather than if they were.

There is no simple or conclusive answer to this problem. All we can do is bring together various pieces of evidence which collectively suggest that our present methods, in spite of their seeming crudity, do give a reasonably accurate picture of the structure of the living cell.

One possibility is to vary the preparative technique. The range of substances which can be used as fixatives is not large, but osmium tetroxide is chemically very different from the aldehydes, and among the latter it is possible to use simple examples, such as formaldehyde, or more complex ones such as glutaraldehyde. On the whole, these substances yield substantially similar results, though some cell components are better preserved by one type of fixative than another. Similarly, variation in the substance used for dehydration, or in the type of plastic used for embedding, makes little difference. It is also possible, for example, to use certain plastics which are water-miscible, so that cells need not be exposed to organic liquids such as ethanol during dehydration. Again, this does not substantially alter the appearance of the resulting micrographs. With regard to staining, it seems that although different salts of heavy metals differ in their affinities for particular cell components, this is a matter of revealing different parts of the cells, rather than of creating artefacts. In general, the preparative methods can be varied over a moderately wide range of conditions without significantly affecting the final picture, and this makes it unlikely that the structures observed are purely artefactual. Nevertheless, it is still the case that if we use almost anything except osmium tetroxide or the aldehydes as fixatives the results are likely to be very different indeed, and much of the fine structure is then no longer visible. It is therefore still necessary to explain why the sort of appearance produced by the 'good' fixatives is regarded as more 'real' than that produced by the others.

A valuable source of evidence here comes from the use of an alternative method of preparation which avoids the use of fixatives altogether. This is the process of freeze-drying. In this method tissues are frozen very rapidly by immersing them in liquid nitrogen. The water is then sublimed off in a vacuum, and when dry the tissue is embedded and sectioned in the usual way. The method has too many disadvantages for routine use (for example cells may be damaged by the formation of ice crystals), but it provides a valuable control for studying the effects of fixatives. It is found that, to a considerable extent, cells prepared by freeze-drying are similar in fine structure to ones which have been exposed to fixatives.

In some cases it has been possible to supplement observations on fixed and sectioned cells by studying fragments of unfixed cells, produced by breaking them up under mild conditions. The tubules seen in sectioned cilia, for example, can be obtained intact by treating cilia with a detergent, which removes their membranes, and they can then be studied by the negative-staining method used for viruses. The shape and dimensions of the tubules in material prepared by this method agree closely with what is seen in sections.

Another source of evidence is available in certain special cases in which it is possible to obtain information about fine structure by indirect methods. An example is the sheath of myelinated nerve fibres. Examination of these with the polarizing microscope, together with observations on the effects of extracting them with solvents, suggested that they must have a layered structure, below the limit of resolution of the light microscope, and this was seen directly when sections of fixed material were examined in the electron microscope. Furthermore, the spacing of the layers can also be determined indirectly, by use of X-ray diffraction, and again the results are in good agreement with those obtained by electron microscopy. The fact that such comparisons of the results of using direct and indirect methods can be made in only a few special cases does not reduce their importance, for if it can be shown even once that the usual preparative techniques used in electron microscopy can yield reliable results, they can be applied with more confidence in other cases.

Finally, there is a type of evidence which is perhaps convincing only to electron microscopists, but which is nevertheless important in practice. This is simply the recognition of the fact that the structures seen in electronmicrographs are often too constant in appearance, too complex, and too asymmetrical to be produced by the action of fixatives, or by dehydration, etc. The arrangement of tubules in a cilium, for example, is not something which we have any reason to suppose could be produced simply by the action of a fixative on a previously homogeneous material. There must have been some antecedent structure closely resembling, even if not actually identical to, the structure observed. This argument can be applied quite widely.

Beyond this we cannot go at present, and it will not be possible to do so until it becomes possible to examine living cells in the electron microscope at a resolution comparable to that which can at present be obtained with sections. Electron microscopists are in general inclined to accept what they see in their micrographs as giving a reliable indication of the structure that existed in the living cells, and the possibility that the present techniques are yielding wholly misleading results can certainly be ruled out. Nevertheless, it must be made clear that almost all electronmicrographs display some features whose validity is uncertain, and which may be artefacts or may be 'real'. Also, it should be remembered that while we may be reasonably confident about accepting the validity of what we do see, we have no reason to be equally sure that we are at present seeing all that there is to be seen. Indeed, it is certain that as preparative methods are improved new aspects of cell structure will become visible.

3.4   Cell structure and cell function

Seeing the intricate fine structure of cells revealed by the electron microscope prompts the obvious question: 'What is it all for?' It can be assumed, as a working hypothesis, that this elaborate organization has some functional significance, and that each component has some role to play in the life of the cell, just as each of the parts of a watch has some function in the whole mechanism. How far, then, can cell fine structure be interpreted in functional terms?

It can be said straight away that electronmicrographs are not in themselves particularly helpful in answering this question. One reason for this should be clear already. A cell is a living, three-dimensional object, which changes with time, while an electronmicrograph is a picture of a static, essentially two-dimensional section. It shows the appearance of one plane through the cell, stopped at a particular moment in time. It is true that by cutting and examining sufficient sections it is possible to build up a three-dimensional model, but even if this is done it remains the image of a dead cell. At present cells can be examined in the electron microscope only after killing them, and it is not possible to study the same cell at different times. Another important limitation of the electron microscope lies in the fact that it tells us very little about the chemistry of cells. It gives information about the distribution of matter in the specimen, but not about the kind of molecules that make up the different structures.

That these are formidable limitations of the electron microscope will be appreciated if we consider the sort of picture of the cell which has come from another type of approach, namely that of the biochemist. For biochemistry is concerned very largely with describing the different types of molecule that exist in cells, and the changes and interactions which they undergo. A great number of different types of molecule occur in any kind of cell, and a great many reactions are going on all the time. These reactions are, almost without exception, catalysed by enzymes. There are, for example, the long processes which result in the liberation of energy from carbohydrates or fats, or there are the elaborate synthetic reactions which result in the formation of proteins or nucleic acids. It is unnecessary to go into details here: we can generalize by saying that for the biochemist the cell is a dynamic chemical machine of great complexity. The electron microscopist cannot offer anything comparable to this: his micrographs are usually uninformative about the chemical nature of different parts of the cell, and they are static.

Yet the balance does not lie wholly with the biochemist. For while he may know a great deal about the chemistry of the cell and be able to trace sequences of reactions in detail, his methods tell him little about the way in which these different processes are organized and interrelated in the cell, and this is an important deficiency. It has become clear that while a great deal can be learnt from investigating and reconstructing biochemical processes in the test tube, a full understanding of cell organization requires much more than this. To start with, by no means all the processes known to go on in living cells can be made to occur in cell extracts. Furthermore, the chemical processes going on in an intact cell are regulated and co-ordinated in ways that cannot be reproduced outside it. When the biochemist grinds up a tissue and studies the processes going on in the resulting suspension he is investigating something a good deal less complicated and less sophisticated in its behaviour than the cells he started with. A cell is not simply a collection of enzymes together with their appropriate substrate molecules.

On the one hand, then, we have a highly detailed picture of the fine structure of cells provided by the electron microscopist, and on the other a highly detailed picture of cell chemistry provided by the biochemist. Neither is a complete description of the cell by itself. The next step is obvious: the two accounts must be brought together and made to supplement each other. To describe in any detail how this can be done would take us beyond the scope of this booklet, but a brief account must be given, since it will help to make clear the sort of role which the electron microscope now plays in the study of cell organization.

The principal method which has been used to investigate the functions of different organelles is that of *cell fractionation*. As its name implies, this technique aims at the separation of the different components of the cell from each other, so that they can then be subjected to biochemical examination. The method is best applied to a tissue such as liver or pancreas, which consists largely of one type of cell. It is illustrated schematically in Fig. 3–6. The tissue is removed rapidly from the animal, cut or minced into small pieces and then broken up by grinding in a glass vessel with a tightly fitting plunger (this is called a homogenizer). The resulting suspension of cell fragments is then centrifuged. At first this is done slowly, so that unbroken cells and other large pieces of debris are spun down, together with nuclei. The supernatant is removed and centrifuged more rapidly, giving a second pellet. The supernatant is again removed and recentrifuged at still higher speed, yielding a further pellet. The appropriate speeds and times used in this procedure have, of course, to be determined by trial and error, but it is not difficult to find conditions under which each of the pellets consists largely of a different type of cell component. Nuclei are sedimented first, as already noted, and they can be freed of larger pieces of cell debris by resuspending the pellet and centrifuging again. It is possible to obtain a pure nuclear fraction in this way. The second pellet, thrown down at intermediate speeds, is seen to consist of small granules when examined in the light microscope, while the final pellet usually has no recognizable structure, its constituents being below the limit of resolution of the light microscope. The nature of these two fractions can be established by electron microscopy: samples of the pellets are fixed, just as if they were pieces of tissue, and then embedded and sectioned as already described. Examination then shows that the granules of the second pellet are mainly mitochondria (identifiable by their characteristic structure, described above), while the final pellet consists mainly of broken fragments of endoplasmic reticulum,

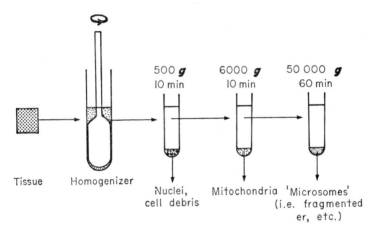

500 g 10 min   6000 g 10 min   50 000 g 60 min

Tissue   Homogenizer

Nuclei, cell debris   Mitochondria   'Microsomes' (i.e. fragmented er, etc.)

Fig. 3–6   Diagram showing the main steps in cell fractionation.

together with ribosomes, which may or may not be attached to the membranes. The exact nature of this final pellet depends on the kind of tissue used as starting material: some cells have more rough-surfaced er than smooth, and so on. The use of the electron microscope is valuable in establishing just what each fraction consists of, and also for determining how far the fractions are pure—that is, consist of only one type of cytoplasmic organelle. Once the identity of the various fractions has been established they can be examined biochemically, their enzyme activities determined, and so on.

With this technique it has been shown that there is a sharp localization of different types of activity in different organelles. The mitochondria, for example, are the site of most of the energy-yielding reactions in which carbohydrates are broken down, with formation of adenosine triphosphate (ATP), while the rough-surfaced er is the chief site of protein synthesis. From plant cells it is possible to obtain a chloroplast fraction and show that it is the site of the photosynthetic reactions. It is unnecessary to enter into details here. The essential point is that each type of structure seen in the electron microscope appears to have a distinctive biochemical function, and carries the enzymes and other equipment necessary to carry it out.

There are other techniques which supplement the fractionation method as a means of studying the localization of function in cells. It is, for example, possible to apply specific staining procedures (analogous to the 'spot tests' used in analytical chemistry) to cells or sections of them, and thereby reveal the presence of certain chemicals or enzymes in particular organelles. This is most commonly done at the light-microscope level, but the methods are gradually being adapted for the electron microscope. The difficulties here lie in devising tests which yield a reaction product visible in the electron microscope (i.e. one which is dense to electrons), and which appears at the original site in the cell of the chemical being studied, and not somewhere else. These are formidable requirements, but it has been possible to localize certain enzymes with a fair degree of precision by such methods.

From what has been said it should be evident that our present understanding of cell organization is the outcome of bringing together a variety of techniques and ideas. The electron microscope reveals enormously more in the way of structure than does any other instrument, yet

---

Plate 3. (a) A mitochondrion, longitudinally sectioned. Note the two external membranes, the inner of which gives rise to the cristae (arrow). Magnification × 60,000.

(b) Cross-section of a flagellum from a protozoon. The main feature is a ring of nine double outer tubules, with a central pair of single ones. All the tubules appear to be hollow. Externally there is a membrane (arrow) which, at the base of the flagellum (not shown here), is continuous with the general plasma membrane of the cell. Magnification × 100,000.

(c) Cross-section of cilia from the gills of a freshwater mussel, *Anodonta*. Note the great similarity to the flagellum in (b). Magnification × 100,000.

(d) Section of an insect flight muscle, showing four myofibrils. Between them are rows of mitochondria, for the most part regularly arranged so that there are two to each sarcomere. The A-bands make up most of the sarcomere length in this muscle, the lighter I-bands (on either side of the Z-lines) being short and relatively inconspicuous. The A-bands are divided into two by a central H-zone. The filaments which make up the myofibrils (see Fig. 3–7) can just be made out. Magnification × 25,000.

(e) Section of a choloroplast from a spinnach leaf. For a description of the structure, see p. 36. Note at the bottom of the picture the cell wall and plasma membrane. Magnification × 32,000.

(a kindly supplied by DR. A. M. MULLINGER; c by DR. I. R. GIBBONS; d by DR. D. S. SMITH; and e by DR. A. D. GREENWOOD.)

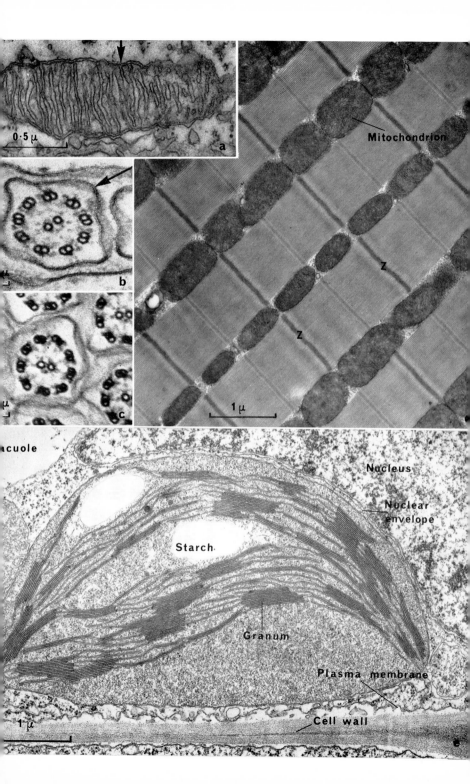

0·5 μ

a

b

c

Mitochondrion

Z

Z

1 μ

Vacuole

Nucleus

Nuclear envelope

Starch

Granum

Plasma membrane

1 μ

Cell wall

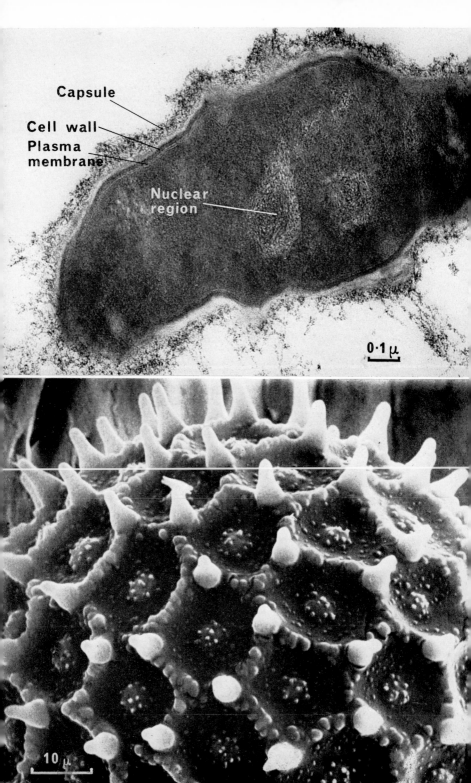

Capsule

Cell wall

Plasma membrane

Nuclear region

0.1 μ

10 μ

in itself it is of extremely limited value as a means of revealing the functional significance of that structure. For that it is necessary to combine its findings with those provided by other methods.

3.5 Bacterial cells

The cells considered in the previous section were those of plants and animals, including protozoa. All of these are built on essentially the same pattern. It is instructive to compare their organization with that of bacteria, which are, of course, very much smaller and notably different in structure.

A typical bacterium is a sphere or rod, usually less than a micron in diameter. Many of them are just on the limit of resolution of the light microscope. They are, in other words, of the same order of size as a mitochondrion. Plate 4a is an electronmicrograph of a section of a bacterium. It is obvious that internally there is nothing comparable with the elaborate fine structure of the cell shown in Plate 2b. The only internal differentiations are the lighter areas containing fine filaments, which are the nuclear regions, containing the bacterial chromosomes. These regions are not bounded by nuclear envelopes but lie freely in the general cytoplasm. Apart from this, the cell contents appear uniform. At higher magnification it would be possible to see that the cell contents are not actually homogeneous, but include, for example, large numbers of ribosomes. These, however, are not attached to membranes but lie freely in the cytoplasm. Other granules of reserve materials may also sometimes be present. There are no elaborate systems of membranes of different kinds; mitochondria, chloroplasts, Golgi bodies, endoplasmic reticulum are all absent. The flagella, if they are present, are simple protein fibres (already described, see p. 24 and Plate 1d), and not the elaborate organelles of plant and animal cells (Plate 3b).

The only prominent cytoplasmic structure in bacteria is the *plasma membrane*, which, as in higher cells, forms the permeability barrier at the cell surface (Plate 4a). Outside this there is a *cell wall*, which is a strong, resistant layer that protects the cell from damage and distortion, while outside this again there may, in some bacteria, be a loose layer of a mucilaginous type of substance which forms the *capsule*.

The smallness of bacteria is therefore correlated with structural simplicity, and this suggests that the great complexity of structure in plant and

---

Plate 4. (a) Section of a bacterium. Note the absence of cytoplasm differentiation. For further explanation, see above. Magnification × 100,000.

(b) Part of a pollen grain (*Ipomoea*) as seen in the scanning electron microscope. Magnification × 2,000.

(a kindly supplied by MISS A. M. GLAUERT from *Br. med. Bull.*, 1962, **18**, no. 3, Plate 30A; b kindly supplied by DR. P. ECHLIN.)

animal cells may be a necessary consequence of their large size. It seems likely that a bacterial type of cell would not work efficiently above a certain size. It becomes essential in a larger cell to separate different sorts of activities and concentrate them in different organelles. In these it is possible to maintain higher rates of reactions than would otherwise be possible, since the enzyme and substrate molecules involved can be kept at higher concentration in a restricted region (such as the inside of a mitochondrion). In some cases it is likely that the enzymes which catalyse the successive steps in a long reaction chain are actually built into the membranes of organelles, so that substrate molecules may pass directly from one to another, instead of having to rely on chance encounters resulting from diffusion in solution. The evidence for such suggestions lies far outside the scope of this booklet, and has in any case not been obtained by use of the electron microscope. The ideas have been introduced here simply in order to indicate that the separation of activities into different organelles was probably a necessity, from the point of view of efficiency, in the evolution of plant and animal cells.

3.6   Specializations in cells and tissues

So far in this chapter we have been concerned only with the basic features of cell fine structure and the methods used to study them. These features can be found in some form in almost all cells, but the relative extent of their development varies a good deal. It has already been noted, for example, that cells which are specialized for secreting protein contain much rough-surfaced endoplasmic reticulum, as well as one or more prominent Golgi bodies. These organelles may occupy much of the cytoplasm of such cells (see Fig. 3–4). As an example of a more extreme type of specialization we shall now consider briefly the fine structure of the striated muscles of vertebrate and invertebrate animals.

Striated muscles consist of cylindrical muscle *fibres*, which are between 10 and 100 μ in diameter and up to several millimeters long. Each fibre is a syncytium, formed by fusion of numerous cells which lose their individuality. Their nuclei remain, usually located near the surface of the fibre, below the bounding membrane (the sarcolemma). Internally the fibres consist largely of bundles of *myofibrils*, each of which is less than 1 μ in diameter (Fig. 3–7). The myofibrils are capable of contraction when teased out of fresh muscle. In the light microscope they show a characteristic and well-known pattern of cross-banding or striation. Plate 3d, is an electron-micrograph which shows this pattern of banding in more detail. Here it can also be seen that the spaces between the myofibrils are almost entirely filled with rows of mitochondria. It will be noted that these are regularly aligned with respect to the striations of the fibrils, there being two (or sometimes one) to every repeating unit or sarcomere. It will be recalled that mitochondria are largely concerned with ATP formation (p. 43), and their striking abundance in muscle is readily explained by the fact that this is an extremely active tissue, and uses large amounts of energy. The only other noteworthy structure in muscle fibres is an elaborate system of delicate tubules and vesicles which run around and between the fibrils. This is known as the *sarcoplasmic reticulum* and is another form which cytoplasmic membranes may take. The sarcoplasmic reticulum plays important parts in the processes of muscle contraction and relaxation. Striated muscle fibres

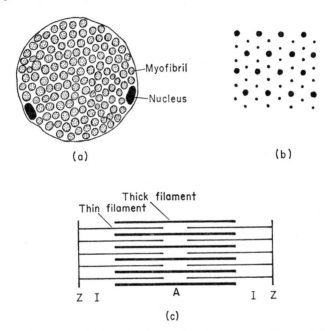

Fig. 3–7   Striated muscle. (a) Transverse section of a single muscle fibre as it appears in the light microscope. Plate 3d shows the myofibrils in a low-magnification electronmicrograph. (b) Arrangement of thick and thin filaments as seen in a small area of a transversely sectioned myofibril examined at high magnification in the electron microscope. (c) Thick and thin filaments, as seen in longitudinal section. During contraction the thin filaments slide further into the A-band.

then, are clearly highly specialized structures, with a complement of organelles designed purely to carry out their function of generating tension in a lengthwise direction.

The electron microscope, in addition to greatly clarifying the structure of the fibres as a whole, has provided important information about the mechanism of muscular contraction. The significant structures here are the myofibrils, and the first point to note about them is that they are made up of longitudinally running filaments. These can be seen even in Plate 3d, which is a comparatively low-magnification micrograph. In high-resolution pictures it can be seen that the filaments are of two kinds, thick and thin, arranged in a regular and characteristic pattern. This is shown diagrammatically in Fig. 3–7. The thick filaments are confined to the A-band of each sarcomere, while the thin ones start in the I-band, where they occur by themselves, but also extend some way into the A-band. Examination of sections of muscle fixed when either contracted or relaxed shows that the extent of overlap of the two kinds of filament in the A-band increases as the muscle contracts. In other words, as the muscle shortens the thin filaments slide further into the A-band. This is the basis of the 'sliding filament' theory of muscle contraction, which proposes that neither type of filament shortens when the muscle contracts but that they simply move relative to each other. In agreement with this theory is the fact that the length of the A-band remains unaltered in contracted muscle, while the I-bands shorten. The electron microscope has provided a good deal of information about the detailed structure of the two types of filament, extending almost down to molecular dimensions, but the exact mechanism by which they are caused to move relative to one another is still not understood.

Striated muscles provide a good example of the way in which the electron microscope has contributed to a considerable and rapid advance in understanding of the structure and function of a familiar type of tissue. Space does not permit other examples to be considered here, but we may generalize by saying that the electron microscope is currently being used to explore the whole range of plant and animal tissues, providing detailed information about the structure and inter-relationships of their constituent cells and yielding many clues as to the way in which these tissues operate. The older accounts of histology based on light microscopy—whether of nervous systems or sense organs, kidney or skin, xylem vessels or meristems—are being supplemented or replaced by more detailed descriptions based on the use of the electron microscope. It may be expected that this will continue to be an active and fruitful field of research in the next few years, and that it will yield particularly valuable results in those cases where the new information about structure can be combined with knowledge of physiology and biochemistry.

Surfaces

4.1 Replicas

In studying certain problems in biology it is necessary to examine surface structures in great detail. The insect cuticle is one example, where knowledge of the structure of the outermost layer may help to explain how insects are waterproofed, and also suggest how insecticides can be designed to penetrate more effectively. With the ordinary type of electron microscope it is not possible to examine such surfaces directly, since an object as large as an insect, or even a piece of insect cuticle, is completely opaque to the electron beam. Only with objects as small as viruses or parts of cells is it

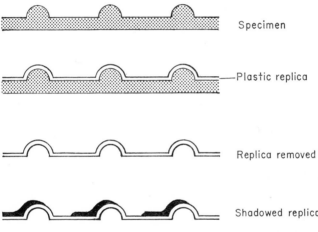

Fig. 4-1   Preparation of a replica.

possible to examine surface detail directly, usually with the help of shadowing or negative staining to enhance contrast. With larger objects it is necessary either to use a different type of electron microscope, which does not depend on transmission of the beam through the specimen (see § 1.8 p. 13) or to adopt an indirect approach. The latter will be described first.

The indirect approach to examining a surface involves making a *replica* of it. That is, a copy of it is prepared which is sufficiently thin and transparent to permit direct examination. There are many ways of making replicas. A simple method is illustrated in Fig. 4-1. In this the specimen is first flooded with a dilute solution of a plastic, such as celloidin, most of which is then drained off. The remaining film is then allowed to dry by evaporation, resulting in an exact copy of the surface in plastic, which can be carefully stripped off. Such a replica may be shadowed (p. 20), in order to enhance contrast and to bring out the three-dimensional structure of the original surface. In the case of leaf and insect surfaces the method just described would not always be suitable, since their outermost layers are waxy and would be likely to dissolve in the plastic solution. This difficulty is overcome by making the initial replica not of plastic but of a thin layer of metal or carbon evaporated on to the surface in a vacuum. Such a replica is too fragile to be stripped immediately from the original, but it can be strengthened temporarily by backing it with a layer of plastic, which is subsequently dissolved away. The replica can again be shadowed. Many modifications of these methods have been devised for the examination of surfaces of various shapes and composition. The basic principle in all of them, however, is that a copy is made of the surface sufficiently thin and transparent to be examined directly in the electron microscope.

4.2   Plant and animal surfaces

A great range of different types of surfaces, both plant and animal, have been examined in the electron microscope. As might be expected, the results provide much new information about details of structure, but they do not lend themselves to convenient summary, and they do not lead to any striking generalizations of the type possible in connection with cell fine structure, where the electron microscope has revealed a common pattern of organization in all plant and animal cells. All that will be done here is to pick out a few examples of the kinds of results which have been obtained.

Among plants several studies have been made of the wax layers on *leaf surfaces*. These prove to have a characteristic sculptured form, which permits the layer to be readily recognized. This is useful in experimental studies on the action of weed-killers and systemic insecticides, the penetration of which into the plant may be facilitated by damage to the wax layer. (A similar situation is found in the case of the wax which forms an outer layer on some insect cuticles: damage by abrasives increases the effectiveness of some insecticides.) *Pollen grains* have been extensively studied and a great deal of new information has been obtained about the minute sculpturing of their surfaces. The earlier studies were made using replicas, prepared by the techniques described above, but more recently the scanning electron microscope has come into use and permits direct examination of pollen grains (and many other objects), almost without preparation. The method of operation of the scanning microscope was very briefly described earlier (p. 14). It has the great advantage that it permits the surfaces of quite large objects to be examined and has a considerable depth of focus, instead of being restricted to a limited focal plane as is the case with the ordinary transmission electron microscope. Plate 4b is an example of the sort of picture of pollen which it yields. It will be seen that the surface bears a regular pattern of ridges and hollows, which are themselves decorated with smaller protuberances and pits. The manner in which this elaborately sculptured surface forms during development of the pollen grain is not yet clear, though some features of the process have been worked out from a study of thin sections in the electron microscope. Much of the interest of such work lies in the fact that an analysis of the mechanism of pattern development in a regular structure such as this may ultimately lead to a better understanding of the way in which the shapes of more complicated structures are generated in cells generally. The detailed morphology of pollen grains is of interest for another reason also, in that when sufficient genera have been studied a comparative analysis of the results may help to throw light on the taxonomic relationships of different families of plants.

The siliceous shells of diatoms are comparable to pollen grains in being elaborately and symmetrically patterned, and they, too, have been extensively studied with the electron microscope. Many small pigmented flagellates and also some of the Foraminifera have elaborate shells or scales of characteristic structure. These are of particular interest since they may become fossilized and provide a valuable means of characterizing and 'zoning' geological deposits. Electron microscopes are being used quite extensively for work of this kind.

Another field of research in which replica techniques (as well as sectioning) have played an important part is the study of plant *cell walls* (Plate 3e). It is possible in these to resolve the basic cellulose microfibrils as long threads, about 100 Å in diameter. The fibrils are often arranged parallel to one another in layers, and the orientation of these with respect to the cell axes can be observed. It is also possible to follow the processes of thickening which go on in, for example, xylem and phloem elements.

In zoology perhaps the chief use which has been made of the electron microscope in examining surfaces has been in the study of various insect materials. Observations of the wax layers of the cuticle have already been mentioned. Studies have also been made of such things as egg surfaces, butterfly scales and cuticular specializations associated with sense organs. The resolving power achieved in such work, especially with the scanning microscope, may be only a few hundred Ångström units, but this is a sufficient improvement on the performance of the light microscope to yield a large amount of new information, which can sometimes be correlated with physiological data.

5 Conclusions

The aim of this booklet has been to give a sketch of how the electron microscope works, and to indicate briefly what sort of results it has yielded in biology. It may perhaps be useful to end by drawing a few general conclusions.

It is clear, in the first place, that the modern electron microscope is a tool of extraordinary refinement. It has opened up to direct inspection a level of structure which previously could be explored only by indirect and frequently imprecise techniques. The observations which have been made on the structure of viruses and on the fine structure of cells and tissues are among the most exciting and important recent discoveries in biology. Studies with the electron microscope have gone far towards closing the gap between the level of large molecules—proteins, polysaccharides, nucleic acids—and the level of structure revealed by the light microscope, and by so doing they have helped greatly to unify our understanding of cell organization.

It is worth pointing out that this important and rapid advance in knowledge has been made possible simply by an advance in *technique*. It did not call for creative intellectual effort comparable to that which led, for example, to the theory of evolution or the chromosome theory of heredity. Given the electron microscope, its application in biology required only the development of suitable preparative methods, and this proved relatively simple. This is not an uncommon phenomenon in the biological sciences. A hundred years ago the development and application of the light microscope was resulting in a comparable and equally important advance in understanding of the organization of cells and tissues. More recently, progress in biochemistry has been very closely dependent on the development of chromatographic techniques and the introduction of radioactive isotopes.

It is important to recognize (and has repeatedly been stressed in this booklet) that, like any other instrument, the electron microscope has its limitations. At present it can be used to examine only dead specimens, usually subjected to the rigours of fixation, embedding, sectioning, and so on. It yields very little in the way of chemical information about the composition of the specimen. These deficiencies have to be made good by the use of other techniques, as outlined in Chapter 3. In the early days of its application in cytology the electron microscope was used with the simple object of accumulating information about the *structure* of cells and tissues. In its essentials, however, much of this work has now been done, and in future the electron microscope will be used as one among many tools for studying cells as dynamic, living organisms, which metabolize, grow, secrete, divide, and so on.

Electron microscopes and the associated techniques are still being actively developed and it is hazardous to try to predict what form they will take in a few years' time. It seems likely, however, that preparative methods will continue to be improved, giving better preservation of structure, and that new techniques will be worked out for studying the chemical composition of specimens. It does not at present seem likely that the resolving power of electron microscopes will be substantially improved in the near future, but it does seem probable that microscopes operating at high accelerating voltages (up to a million volts) will come into increasing use. Just conceivably these may permit limited observations on living cells, by providing a beam powerful enough to penetrate both the cell and the walls of the chamber in which it would have to be housed. Whether cells would survive the intense irradiation long enough to yield useful information remains to be determined.

Further Reading

CASPAR, D. L. D. and KLUG, A. (1962). *Cold Spring Harb. Symp. Quant. Biol.*, **27**, 1 (discussion of virus structure).

COSSLETT, V. E. (1966). *Modern Microscopy.* Bell, London.

FAWCETT, D. W. (1966). *An Atlas of Fine Structure: The Cell.* Saunders, Philadelphia & London.

HAGGIS, G. H. (1966). *The Electron Microscope in Molecular Biology.* Longmans, London.

HAGGIS, G. H., MICHIE, D., MUIR, A. R., ROBERTS, K. B. and WALKER, P. B. M. (1964). *Introduction to Molecular Biology.* Longmans, London.

HUGHES, A. (1959). *A History of Cytology.* Abelard-Schuman, London & New York.

HURRY, S. W. (1965). *The Microstructure of Cells.* Murray, London.

KAY, D. (ed.) (1965). *Techniques for Electron Microscopy*, 2nd ed. Blackwell, Oxford.

LOEWY, A. G. and SIEKEVITZ, P. (1963). *Cell Structure and Function.* Holt, Rinehart and Winston, New York.

PEASE, D. C. (1964). *Histological Techniques for Electron Microscopy*, 2nd ed. Academic Press, New York & London.

PORTER, K. R. and BONNEVILLE, M. A. (1964). *The Fine Structure of Cells and Tissues*, 2nd ed. Lea and Febiger, Philadelphia.

RAMSAY, J. A. (1965). *The Experimental Basis of Modern Biology.* University Press, Cambridge.